REA's Test Prep Bo

(more on next page)

(continued from front page)

" I am writing to congratulate you on preparing an exceptional study guide. In five years of teaching this course, I have never encountered a more thorough, comprehensive, concise and realistic preparation for this examination. "
Teacher, Davie, FL

" I have found your publications, *The Best Test Preparation...*, to be exactly that. "
Teacher, Aptos, CA

" I used your book to prepare for the test and found that the advice and the sample tests were highly relevant... Without using any other material, I earned very high scores and will be going to the graduate school of my choice. "
Student, New Orleans, LA

" I used your *CLEP Introductory Sociology* book and rank it 99% — thank you! "
Student, Jerusalem, Israel

" Your *GMAT* book greatly helped me on the test. Thank you. "
Student, Oxford, OH

" I recently got the *French SAT II* Exam book from REA. I congratulate you on first-rate French practice tests. "
Instructor, Los Angeles, CA

" Your *AP English Literature and Composition* book is most impressive. "
Student, Montgomery, AL

" The REA *LSAT* Test Preparation guide is a winner! "
Instructor, Spartanburg, SC

The Best Test Preparation for the

SAT Subject Test
Math
Level 2

7th EDITION

With REA's TEST*ware*® on CD-ROM

Staff of Research & Education Association

Research & Education Association
Visit our website at
www.rea.com

Research & Education Association
61 Ethel Road West
Piscataway, New Jersey 08854
E-mail: info@rea.com

The Best Test Preparation for the
SAT SUBJECT TEST: MATH LEVEL 2
With TEST*ware*® on CD-ROM

Printed in the United States of America

Library of Congress Control Number 2006904035

International Standard Book Number 0-7386-0264-7

Windows® is a registered trademark of Microsoft Corporation.

REA® and TEST*ware*® are registered trademarks of
Research & Education Association, Inc.

CONTENTS

ABOUT RESEARCH & EDUCATION ASSOCIATION

Founded in 1959, Research & Education Association (REA) is dedicated to publishing the finest and most effective educational materials—including software, study guides, and test preps—for students in middle school, high school, college, graduate school, and beyond.

REA's Test Preparation series includes books and software for all academic levels in almost all disciplines. Research & Education Association publishes test preps for students who have not yet entered high school, as well as high school students preparing to enter college. Students from countries around the world seeking to attend college in the United States will find the assistance they need in REA's publications. For college students seeking advanced degrees, REA publishes test preps for many major graduate school admission examinations in a wide variety of disciplines, including engineering, law, and medicine. Students at every level, in every field, with every ambition can find what they are looking for among REA's publications.

REA presents tests that accurately depict the official exams in both degree of difficulty and types of questions. REA's practice tests are always based upon the most recently administered exams, and include every type of question that can be expected on the actual exams.

REA's publications and educational materials are highly regarded and continually receive an unprecedented amount of praise from professionals, instructors, librarians, parents, and students. Our authors are as diverse as the fields represented in the books we publish. They are well known in their respective disciplines and serve on the faculties of prestigious high schools, colleges, and universities throughout the United States and Canada.

We invite you to visit us at *www.rea.com* to find out how "REA is making the world smarter."

STAFF ACKNOWLEDGMENTS

We would like to thank Larry B. Kling, Vice President, Editorial, for supervising development; Pam Weston, Vice President, Publishing, for setting the quality standards for production integrity and managing the publication to completion; John Paul Cording, Vice President, Technology, for coordinating the design and development of REA's TEST*ware*® software; Diane Goldschmidt, Associate Editor, for post-production quality assurance; Dr. Kai Miao for his editorial contributions; Heena Patel and Michelle Boykins, Technology Project Managers, for their software design contributions and testing efforts; and Network Typesetting, Inc., for typesetting the manuscript. Special thanks are also extended to Christine Saul, Senior Graphic Artist, for designing our cover.

THE SAT SUBJECT TEST IN

Math
Level 2

CHAPTER 1

Passing the SAT Math Level 2 Subject Test

Chapter 1

PASSING THE
SAT MATH LEVEL 2 SUBJECT TEST

ABOUT THIS BOOK AND TEST*ware*®

This book, along with REA's exclusive TEST*ware*® software, provides you with an accurate and complete representation of the SAT Mathematics Level 2 Subject Test. Inside you'll find a complete course review that presents no-nonsense, relevant information, along with clear-eyed strategies you need to do well on the exam. But we don't stop there: you'll also find six full-length REA practice tests based on official exam questions released by the College Entrance Examination Board. REA's practice tests contain every type of question that you can expect to encounter on the SAT Math Level 2 Subject Test. Following each of our practice tests, you will find an answer key with detailed explanations designed to help you master the test material.

Practice Tests 1 and 2 are also included on the enclosed TEST*ware*® CD. The software provides the benefits of instantaneous, accurate scoring and enforced time conditions.

ABOUT THE TEST

Who takes the test and what is it used for?

Planning to go to college? Then you should take the SAT Math Level 2 Subject Test in *either* of these cases:

(1) Any of the colleges to which you're applying require the test for admission.

(2) You wish to demonstrate proficiency in mathematics.

The SAT Math Level 2 exam is designed for students who have taken more than three years of college preparatory mathematics, including two years of algebra, one year of geometry, and some pre-calculus and/or calculus.

Who administers the test?

The SAT Math Level 2 Subject Test is developed by the College Board and administered by Educational Testing Service (ETS). The test development process involves the assistance of educators throughout the country, and is designed and implemented to ensure that the content and difficulty level of the test are appropriate.

When should the SAT Math Level 2 Subject Test be taken?

If you are applying to a college that requires Subject Test scores as part of the admissions process, you should take the SAT Math Level 2 Subject Test by November or January of your senior year. If your scores are being used only for placement purposes, you may be able to take the test in the spring. Contact the colleges to which you are applying for details.

When and where is the test given?

The SAT Math Level 2 Subject Test is offered six times a year at many locations—mostly high schools—throughout the country. The test is given in October, November, December, January, May, and June.

To receive information on upcoming administrations of the exam, consult your guidance counselor or contact:

College Board SAT Program
P.O. Box 6200
Princeton, NJ 08541-6200
Phone: (609) 771-7600
Website: http://www.collegeboard.com

Is there a registration fee?

You must pay a registration fee to take the SAT Math Level 2. Consult the College Board Website (*www.collegeboard.com*) for information on the fee structure. Financial assistance may be granted in certain situations. To find out if you qualify and to register for assistance, contact your academic advisor.

What kind of calculator can I use?

If at all possible, bring a graphing calculator on test day. The test assumes that most students use a graphing calculator, and having one at your side may give you an edge. Consult official ETS and College Board publications (including Collegeboard.com) for more specifics. No pocket organizers, hand-held minicomputers, paper tape, or noisy calculators may be used. In addition, no calculator requiring an external power source will be allowed and calculators may not be shared—you must bring your own.

Thoroughly acquaint yourself with the operation of your calculator. Your performance could suffer if, say, you spend too much time searching for the correct function, or fail to keep in mind that the test's answer choices are rounded, or forget to switch to the correct calculating mode.

HOW TO USE THIS BOOK AND TEST*ware*®

What do I study first?

Because the SAT Math Level 2 Subject Test is designed to test knowledge that has been acquired throughout your education, the key to solid preparation is to thoroughly review the subject matter. Refresh yourself by studying our review material and taking the sample tests provided in this book. Our practice tests will expose you to the types of questions, directions, and format that are characteristic of the SAT Math Level 2 Subject Test.

To begin your studies, read over the reviews and the suggestions for test-taking, take Practice Test 1 on CD-ROM to determine your area(s) of weakness, and then restudy the review material, focusing on your specific problem areas. The course review includes the information you need to know when taking the exam. Make sure to follow up your diagnostic work by taking Practice Test 2 on CD-ROM to become familiar with the format and feel of the SAT Math Level 2 Subject Test.

When should I start studying?

It is never too early to start studying for the SAT Math Level 2 test. The earlier you begin, the more time you will have to sharpen your skills. Do not procrastinate! Cramming is *not* an effective way to study, since it does not allow you the time needed to learn the test material. The sooner you learn the format of the exam, the more comfortable you will be when you take it.

FORMAT OF THE SAT MATH LEVEL 2

The Math Level 2 Subject Test is a one-hour exam consisting of 50 multiple-choice questions. Each question has five possible answer choices, lettered (A) through (E).

What's on the test?

Here's the approximate distribution of topics covered on the exam:

Topic	Percentage	Number of Questions
Algebra	18	9
Geometry	20	10
• Three-dimensional	8	4
• Coordinate	12	6
Trigonometry	20	10
Functions	24	12
Statistics/Probability	6	3
Miscellaneous*	12	6

* Includes logic and proof, elementary number theory, sequences, and limits.

Questions on the test are also grouped according to whether or not you need to use your calculator:

Category	Definition	Approximate Percentage of Questions
Calculator inactive	Calculator not necessary or advantageous	40
Calculator neutral	Calculator may be useful, but not absolutely necessary	} 60
Calculator active	Calculator is necessary to solve the problem	

SCORING THE SAT MATH LEVEL 2

How do I score my practice tests?

The SAT Math Level 2 Test is scored on a 200-800 scale. As with the actual test, your practice tests are scored by crediting one point for each correct answer and deducting one-fourth of a point for each incorrect answer. There is no deduction for answers that you skip. Use the worksheet below to calculate your raw score and to record your scores for the six practice tests. The table on the following page shows you how to convert your raw score to a scaled score.

PRACTICE-TEST SCORING WORKSHEET

_____   −   (_____ x 1/4)   =   _____

number correct

number incorrect
(do not include
unanswered questions)

Raw Score
(round to nearest
whole number)

	Raw Score	**Scaled Score**
Test 1	_____	_____
Test 2	_____	_____
Test 3	_____	_____
Test 4	_____	_____
Test 5	_____	_____
Test 6	_____	_____

PRACTICE-TEST SCORE CONVERSION TABLE*

Raw Score	Scaled Score	Raw Score	Scaled Score
50	800	18	560
49	790	17	560
48	790	16	550
47	780	15	540
46	780	14	530
45	770	13	530
44	770	12	520
43	760	11	510
42	760	10	500
41	750	9	500
40	750	8	490
39	740	7	480
38	740	6	480
37	730	5	470
36	730	4	460
35	720	3	450
34	710	2	440
33	700	1	430
32	690	0	410
31	680	–1	390
30	670	–2	370
29	660	–3	360
28	650	–4	340
27	640	–5	340
26	630	–6	330
25	630	–7	320
24	620	–8	320
23	610	–9	320
22	600	–10	320
21	590	–11	310
20	580	–12	310
19	570		

* Readers are cautioned that the scaled scoring conversions used here *approximate* those for the actual test. Due to statistical formulas used by the test administrator and slight variations from one edition of the test to another, your score on the SAT Mathematics Level 2 Subject Test may be somewhat higher or lower than what you achieve on REA's practice tests in this book.

STUDYING FOR THE SAT MATH LEVEL 2

It is critical to choose the time and place for studying that works best for you. Some students may set aside a certain number of hours every morning to study, while others may choose to study at night before going to sleep. Only you can determine when and where your study time will be most effective. Be consistent and use your time wisely. Work out a study routine and stick to it!

When you take the practice tests, try to make your testing conditions as realistic as possible. Turn your television and radio off, and sit down at a quiet table free from distraction. Be sure to clock yourself with a timer.

As you complete each practice test, score your test and thoroughly review the explanations for the questions you answered incorrectly; however, do not review too much at any one time. Concentrate on one problem area at a time by reviewing the questions and explanations, and by studying our review until you are confident you have mastered the material.

Keep track of your scores. By doing so, you will be able to gauge your progress and discover weaknesses in particular sections. You should carefully study the reviews that cover your areas of difficulty, as this will build your skills in those areas.

TEST-TAKING TIPS

Although you may be unfamiliar with standardized tests such as the SAT Math Level 2 Subject Test, there are many ways to acquaint yourself with this type of examination and help alleviate any test-taking anxieties. Listed below are ways to help you become accustomed to the SAT Math Level 2 Subject Test, some of which may apply to other standardized tests as well.

Become comfortable with the format of the exam. When you are practicing to take the SAT Math Level 2 Subject Test, simulate the conditions under which you will be taking the actual test. Stay calm and pace yourself. After simulating the test only a couple of times, you will boost your chances of doing well, and you will be able to sit down for the actual exam with much more confidence.

Know the directions and format for each section of the test. Familiarizing yourself with the directions and format of the exam will not only save you time, but will also ensure that you are familiar enough with the SAT Math Level 2 Subject Test to avoid anxiety (and the mistakes caused by being anxious).

Do your scratchwork in the margins of the test booklet. You will not be given scrap paper during the exam, and you may not perform scratchwork on your answer sheet. Space is provided in your test booklet to do any necessary work or draw diagrams.

If you are unsure of an answer, guess. If you *do* guess, guess wisely. Use the process of elimination by going through each answer to a question and ruling out as many of the answer choices as possible. By eliminating three answer choices, you give yourself a 50/50 chance of answering correctly since there will only be two choices left from which to make your guess.

Mark your answers in the appropriate spaces on the answer sheet. Each numbered row will contain five ovals corresponding to each answer choice for that question. Fill in the oval that corresponds to your answer darkly, completely, and neatly. You can change your answer, but only after completely erasing the old one. Any stray lines or unnecessary marks may cause the machine to score your answer incorrectly. When you have finished working on a section, you may want to go back and check to make sure your answers correspond to the correct questions. Marking one answer in the wrong space will throw off the rest of your test, whether it is graded by machine or by hand.

You don't have to answer every question. You are not penalized if you do not answer every question. The only penalty you receive is if you answer a question incorrectly. Try to use the guessing strategy, but if you are truly stumped by a question, you do not have to answer it.

Work quickly and steadily. You have a limited amount of time to work on each section, so you need to work quickly and steadily. Avoid becoming ensnared by any one problem. Taking the practice tests in this book will help you to learn how to budget your time.

Before the Test

Make sure you know where your test center is well in advance of your test day so you do not get lost on the way there. On the night before the test, gather together the materials you will need the next day:

- Your admission ticket
- Personal identification, which *must* include a photograph *or* physical description (written in English)[†], your name, and your signature (e.g., driver's license, student identification card, or current alien registration card).
- Two No. 2 pencils with erasers
- Your calculator
- Directions to the test center
- A wristwatch (if you wish) but not one that makes noise, as it may disturb other test-takers

On the day of the test, you should wake up early (it is hoped after a decent night's rest) and have a good breakfast. Dress comfortably, so that you are not distracted by being too hot or too cold while taking the test. Plan to arrive at the test center early. This will allow you to collect your thoughts and relax before the test, and will also spare you needless stress. If you arrive after the test begins, you will not be admitted and you will not receive a refund.

---

† **The physical description must be signed in the presence of a school official, who must then co-sign it. Consult your test bulletin for details.**

During the Test

When you arrive at the test center, try to find a seat where you feel you will be comfortable. Follow all the rules and instructions given by the test supervisor. If you do not, you risk being dismissed from the test and having your scores canceled.

Once all the test materials are distributed, the test instructor will give you directions for filling out your answer sheet. Fill this sheet out carefully since this information will appear on your score report.

After the Test

When you have completed the SAT Math Level 2 Subject Test, you may hand in your test materials and leave. Then, go home and relax!

You can expect to receive your score report in four to five weeks. The report will include all your scores and details on how to interpret them.

INSTALLING REA's TEST*ware*®

SYSTEM REQUIREMENTS

Pentium 75 MHz (300 MHz recommended) or higher or compatible processor; Microsoft Windows 98 or later; 64 MB Available RAM; Internet Explorer 5.5 or higher

INSTALLATION

1. Insert the SAT Math Level 2 Subject Test TEST*ware*® CD-ROM into the CD-ROM drive.
2. If the installation doesn't begin automatically, from the Start Menu choose the RUN command. When the RUN dialog box appears, type d:\setup (where D is the letter of your CD-ROM drive) at the prompt and click OK.
3. The installation process will begin. A dialog box proposing the directory "Program Files\REA\SAT_Math2" will appear. If the name and location are suitable, click OK. If you wish to specify a different name or location, type it in and click OK.
4. Start the SAT Math Level 2 Subject Test TEST*ware*® application by double-clicking on the icon.

REA's SAT Math Level 2 Subject Test TEST*ware*® is **EASY** to **LEARN AND USE**. To achieve maximum benefits, we recommend that you take a few minutes to go through the on-screen tutorial on your computer. The "screen buttons" are also explained there to familiarize you with the program.

SSD ACCOMMODATIONS FOR STUDENTS WITH DISABILITIES

Many students qualify for extra time to take the SAT Math Level 2 Subject Test, and our TEST*ware*® can be adapted to accommodate your time extension. This allows you to practice under the same extended time accommodations that you will receive on the actual test day. To customize your TEST*ware*® to suit the most common extensions, visit our Website at http://www.rea.com/ssd.

TECHNICAL SUPPORT

REA's TEST*ware*® is backed by customer and technical support. For questions about **installation or operation of your software**, contact us at:

> **Research & Education Association**
> **Phone: (732) 819-8880 (9 a.m. to 5 p.m. ET, Monday–Friday)**
> **Fax: (732) 819-8808**
> **Website: http://www.rea.com**
> **E-mail: info@rea.com**

Note to Windows XP Users: In order for the TEST*ware*® to function properly, please install and run the application under the same computer-administrator level user account. Installing the TEST*ware*® as one user and running it as another could cause file-access path conflicts.

THE SAT SUBJECT TEST IN

Math
Level 2

CHAPTER 2

Course Review

Chapter 2

COURSE REVIEW

ALGEBRA

Absolute Value

$$|a| = \begin{cases} a \text{ if } a > 0 \\ -a \text{ if } a < 0 \end{cases}$$

Equivalently, the distance on the number line from a to 0 is the absolute value of a.

Properties of absolute values:

$$|-a| = |a|$$

$|a| \geq 0$, equality holding if and only if $a = 0$

$$\left| \frac{a}{b} \right| = \frac{|a|}{|b|}$$

$$|ab| = |a||b|$$

$$|a|^2 = a^2$$

Algebraic Laws and Operations

To add two numbers with like signs, add their absolute values and prefix the sum with the common sign.

Example:

$$-2 + (-3) = -5$$

To add two numbers with unlike signs, find the difference between their absolute values, and prefix the result with the sign of the number with the greater absolute value.

Example:

$$-5 + 2 = -3$$

To subtract a number b from another number a, change the sign of b and add to a.

Example:

$$-2 - (-3) = -2 + 3 = 1$$

To multiply (or divide) two numbers having like signs, multiply (or divide) their absolute values and prefix the result with a positive sign.

Example:

$$(-2)(-3) = 6$$

To multiply (or divide) two numbers having unlike signs, multiply (or divide) their absolute values and prefix the result with a negative sign.

Example:

$$6 \div (-2) = -3$$

Negative and Fractional Exponents

$$a^0 = 1, \text{ if } a \neq 0$$

$$a^{-n} = \frac{1}{a^n}$$

$$a^{\frac{m}{n}} = \sqrt[n]{a^m}$$

Complex Numbers

A *complex number* is of the form $a + bi$ where a and b are constants, $b \neq 0$, and $i^2 = -1$.

To add, subtract or multiply complex numbers, compute in the usual way, replace i^2 with -1 and simplify.

$$(a + bi) + (c + di) = (a + c) + (b + d)i$$

$$(a + bi) - (c + di) = (a - c) + (b - d)i$$

$$(a + bi)(c + di) = ac + adi + bci + bdi^2 = ac - bd + (ad + bc)i$$

Polynomials and Quadratic Equations

If we complete the square in the quadratic equation:

$$ax^2 + bx + c = 0$$

we find the quadratic formula:

$$x = \frac{-b \pm \sqrt{b^2 - 4ac}}{2a}$$

Every quadratic equation can be solved using the quadratic formula, but if factoring works, it is usually easier.

The expression $b^2 - 4ac$ under the square root symbol is called the *discriminant* of the quadratic equation. If the discriminant is zero, then both solutions (or roots) are the same.

If the discriminant is negative, then the roots are not real numbers; they are complex numbers, because the square root of a negative number is imaginary (that is, it needs to be written in terms of i).

If the discriminant is positive, then there are two different real solutions.

In order to solve the polynomial equation (of degree n)

$$a_0x^n + a_1x^{n-1} + \ldots + a_{n-1}x^1 + a_nx^0 = 0$$

we would like to factor the polynomial on the left of the equality.

In order to find whether $(x - c)$ is a factor of the polynomial, we might divide the polynomial by $(x - c)$. If there is no remainder, then $(x - c)$ is a factor.

Example:

Solve for x:

$$2x^3 - x^2 + 2x - 3 = 0$$

Solution:

After reading property 4 on the next page, you can see how we could guess that 1 might be a solution (a root is a solution) of the equation. In order to find whether 1 is a solution, we may try to factor $(x - 1)$ from the polynomial. We could divide 6 by 2 to find out whether 2 is a factor of 6. Similarly, we may divide $2x^3 - x^2 + 2x - 3$ by $(x - 1)$ to find out whether $x - 1$ is a factor of $2x^3 - x^2 + 2x - 3$.

The division might look like this:

$$
\begin{array}{r}
2x^2 + x + 3 \\
x - 1 \overline{\smash{\big)}\ 2x^3 - x^2 + 2x - 3} \\
\underline{2x^3 - 2x^2} \\
x^2 + 2x \\
\underline{x^2 - x} \\
3x - 3 \\
\underline{3x - 3}
\end{array}
$$

There is no remainder, so

$$2x^3 - x^2 + 2x - 3 = (x - 1)(2x^2 + x + 3)$$

The first factor indicates 1 is a solution. If we use the quadratic formula on the second factor, we find

$$x = \frac{-b \pm \sqrt{b^2 - 4ac}}{2a}$$
$$= \frac{-1 \pm \sqrt{1^2 - 4(2)(3)}}{2(2)}$$
$$= \frac{-1 \pm \sqrt{-23}}{4}$$

So 1 is the only real solution.

If r_1 and r_2 are the two complex solutions given by the quadratic formula, then the quadratic polynomial can be factored as follows:

$$(2x^2 + x + 2) = (x - r_1)(x - r_2)$$

This fact may be checked by multiplying the factors on the right.

Properties of Polynomials:

1. Every polynomial has exactly n factors of the form $(x - r)$. So every polynomial equation of degree n has exactly n roots (solutions). Some of the roots may not be real. Some of the roots may be duplicates of other roots.

2. If a polynomial equation $f(x) = 0$ with real coefficients has a root $a + bi$, then the conjugate of this complex number $a - bi$ is also a root of $f(x) = 0$.

3. If $a + \sqrt{b}$ is a root of the polynomial equation $f(x) = 0$ with rational coefficients, then $\sqrt{b}$ is also a root, where a and b are rational and $\sqrt{b}$ is irrational.

4. If a rational fraction in lowest terms b/c is a root of the equation

    $$a_n x^n + a_{n-1} x^{n-1} + \ldots + a_1 x + a_0 = 0$$

    $a_0 \neq 0$, and the a_i are integers then b is a factor of a_0, and c is a factor of a_n.

Factorial Notation

If n is a positive integer, then

$$n! = n(n-1)(n-2)\ldots1$$

$$0! = 1$$

Example:

$$4! = 4(3)(2)(1) = 24$$

Binomial Theorem

$$\binom{n}{k} = \frac{n!}{(n-k)!\,k!}$$

$\binom{n}{k}$ is also written as $n^C k$

The following equation is the binomial theorem or binomial expansion:

$$(x+y)^n = \sum_{k=0}^{n} \binom{n}{k} x^{n-k} y^k$$

$$= \binom{n}{0} x^n + \binom{n}{1} x^{n-1}y + \binom{n}{2} x^{n-2}y^2 + \ldots + \binom{n}{k} x^{n-k}y^k + \binom{n}{n} y^n$$

Example:

Find the expansion of $(a - 2x)^7$

Solution:

Use the binomial formula:

$$(u + v) = u^n + nu^{n-1}v + \frac{n(n-1)}{2} u^{n-2}v^2 + \frac{n(n-1)(n-2)}{2 \times 3} u^{n-3}v^3 + \ldots + v^n$$

and substitute a for u and $(-2x)$ for v and 7 for n to obtain:

$$(a - 2x)^7 = \left[a + (-2x)\right]^7$$

$$= a^7 + 7a^6(-2x) + \frac{7 \times 6}{2} a^5(-2x)^2 + \frac{7 \times 6 \times 5}{2 \times 3} a^4(-2x)^3$$

$$+ \frac{7 \times 6 \times 5 \times 4}{2 \times 3 \times 4} a^3(-2x)^4 + \frac{7 \times 6 \times 5 \times 4 \times 3}{2 \times 3 \times 4 \times 5} a^2(-2x)^5$$

$$+ \frac{7 \times 6 \times 5 \times 4 \times 3 \times 2}{2 \times 3 \times 4 \times 5 \times 6} a^1(-2x)^6 + \frac{7 \times 6 \times 5 \times 4 \times 3 \times 2 \times 1}{2 \times 3 \times 4 \times 5 \times 6 \times 7} a^0(-2x)^7$$

$$(a - 2x)^7 = a^7 - 14a^6x + 84a^5x^2 - 280a^4x^3 + 560a^3x^4$$

$$-672a^2x^5 + 448ax^6 - 128x^7$$

Pascal's Triangle:

The coefficients of $(a + b)^0$, $(a + b)^1$, $(a + b)^2$, ..., $(a + b)^n$ can be obtained from Pascal's Triangle:

$(a + b)^0$	1
$(a + b)^1$	1 1
$(a + b)^2$	1 2 1
$(a + b)^3$	1 3 3 1
$(a + b)^4$	1 4 6 4 1
$(a + b)^5$	1 5 10 10 5 1
$(a + b)^6$	1 6 15 20 15 6 1
$(a + b)^7$	1 7 21 35 35 21 7 1
$(a + b)^8$	1 8 28 56 70 56 28 8 1
$(a + b)^9$	1 9 36 84 126 126 84 36 9 1

where each number in the triangle is the sum of the two numbers above it, or one if it is on the edge.

SOLID GEOMETRY

Cubes, Cylinders, and Pyramids

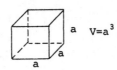

Cube

The volume of a cube with edge a is

$$V = a^3$$

The surface area of a cube with edge a is

$$A = 6a^2$$

Cylinder

The volume of a right circular cylinder with radius r and height h is

$$V = \pi r^2 h$$

The surface area of a right circular cylinder with radius r and height h is

$$A = 2\pi r^2 + 2\pi rh$$

Pyramid

Cone

The volume of a pyramid or cone with base of area A and height h is

$$V = \frac{1}{3}Ah$$

Other Volumes and Surface Areas

If the base of a cone is a circle with radius r, then the area of the circle is

$$A = \pi r^2$$

So, using the formula for the volume of a pyramid or cone, we conclude the volume of a cone with height h base of radius r is

$$V = \frac{1}{3}\pi r^2 h$$

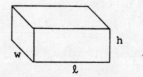

Rectangular solid

The volume of a rectangular solid with length l, width w and height h is

$$V = lwh$$

Sphere

The volume of a sphere with radius r is

$$V = \frac{4}{3}\pi r^3$$

The surface area of a sphere with radius r is

$$A = 4\pi r^2$$

COORDINATE GEOMETRY

Two- or Three-Dimensional Coordinates

Traditionally, a two-dimensional graph is drawn with the positive x-axis extending to the right and the positive y-axis extending up.

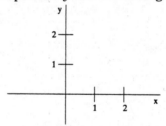

Traditionally, a three-dimensional graph is drawn with the positive x-axis extending to the right, and the positive y-axis extending up. The positive z-axis is pictured as extending toward the reader.

A point on the graph may be pictured by drawing lines to the point parallel to the axes.

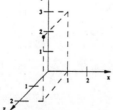

The point (1,2,3)

The equation for a plane is

$$ax + by + cz = d$$

where a, b, c, and d are constants and a, b, and c are not all zero.

The equation for the surface of a sphere with radius r and center (a, b, c) is

$$(x - a)^2 + (y - b)^2 + (z - c)^2 = r^2$$

Transformations

Translations:

Suppose we know the graph of an equation, and we want to find the equation of a congruent graph which is moved h units to the right and k units higher than the original graph. Then we may replace x and y in the original equation with x' and y' respectively in the new equation, where

$x' = x - h$, and

$y' = y - k$

Example:

Given the graph of $y = x^2$, find the equation of a congruent graph with lowest point at $(2,3)$.

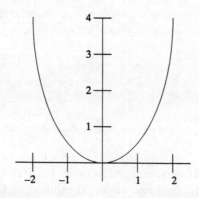

$y = x^2$

Solution:

$(y - 3) = (x - 2)^2$

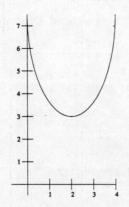

$(y - 3) = (x - 2)^2$

Rotations:

Suppose we know the graph of an equation, and we want to find the equation of a congruent graph which is rotated counterclockwise about the origin through an angle A. Then we may replace x and y in the original equation with x' and y' respectively in the new equation, where

$x' = x \cos \angle A + y \sin \angle A$, and

$y' = -x \sin \angle A + y \cos \angle A$

Example:

Find the equation of the parabola congruent to the parabola given in the previous exercise, rotated 90° counterclockwise about the origin.

Solution:

$y' = (x')^2$

$-x \sin 90° + y \cos 90° = (x \cos 90° + y \sin 90°)^2$

$-x(1) + y(0) = (x(0) + y(1))^2$

$-x = y^2$

$x = -y^2$

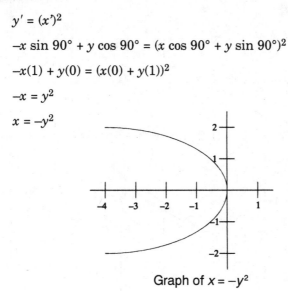

Graph of $x = -y^2$

Vectors

If $M = (2,1)$ and $N = (4, -1)$, then the vector MN can be represented by a straight arrow starting at M and ending at N.

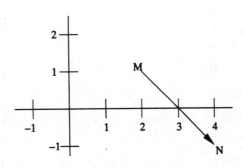

Any other vector with the same length and direction is equal to the vector *MN*.

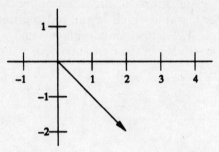

Vector equal to the vector *MN*

A vector that starts from the origin may be named by the terminal point alone. So the vector above,

(2,–2)

is equal to **MN**.

If M, N and P are points, then **MN** + **NP** += **MP**.

Example:

If $M = (2,1)$, $N = (4,2)$ and $P = (5,3)$, find **MN** + **NP**.

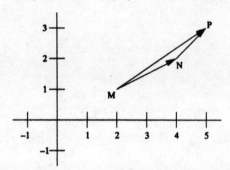

First Solution:

MN + **NP** = (3, 2)

Second Solution:

MN = (4–2, 2–1) = (2, 1)

NP = (5–4, 3–2) = (1, 1)

MN + **NP** may be found by adding the x-coordinates and adding the y-coordinates.

(2 + 1, 1 + 1) = (3, 2)

Vectors are used to represent quantities that have both magnitude and direction. For example, wind velocity and force may be represented with vectors. The length of the vector represents the magnitude of the wind or force.

Polar Coordinates

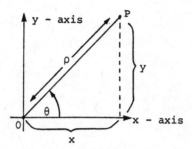

The polar coordinates of a point P which is a distance of ρ from the origin and which lies on the terminal side of an angle θ with initial side on the positive x-axis are ρ and θ.

$$P = (\rho, \theta)$$

The rectangular coordinates of the point P are

$$x = \rho \cos \theta$$

$$y = \rho \sin \theta$$

Given a point (x,y) in rectangular coordinates, the polar coordinates can be found. The angle, θ, is the unique angle for which the above two equations are true, and ρ is given by

$$\rho = \sqrt{x^2 + y^2}$$

Example:

Transform the equation $x^2 + y^2 - x + 3y = 3$ to a polar equation.

Solution:

Substituting for x and y, we get

$$(\rho \cos \theta)^2 + (\rho \sin \theta)^2 - (\rho \cos \theta) + 3(\rho \sin \theta) = 3$$

Factoring, we get

$$\rho^2(\cos^2 \theta + \sin^2 \theta) - \rho(\cos \theta - 3\sin \theta) = 3$$

Replacing $\cos^2 \theta + \sin^2 \theta$ with 1, we get

$$\rho^2 - \rho(\cos \theta - 3\sin \theta) = 3$$

Example:

Transform the equation $\rho = 2 \cos \theta$ to rectangular coordinates.

Solution:

Since $x = \rho \cos \theta$, we have

$$\cos \theta = \frac{x}{\rho} = \frac{x}{\sqrt{x^2 + y^2}}$$

so we may replace $\cos \theta$ in the original equation, $\rho = 2 \cos q$ and also replace ρ to get

$$\sqrt{x^2 + y^2} = \frac{2x}{\sqrt{x^2 + y^2}}$$

Simplifying, we get

$$x^2 + y^2 = 2x$$

Parametric Equations

Equations defining variables in terms of another variable are called *parametric equations*. If coordinates x and y of a point are expressed in terms of time, t, then the parametric equations tell not only where the point (x,y) can be, but also at what time the point is at the location (x,y).

Example:

Graph the point (x,y) given by the equations

$$x = t + 1$$

$$y = t^2 + 2t + 1$$

noting the location of the point at the times when $t = 0$ and when $t = 1$. Solve for y in terms of x.

Solution:

T	X	Y
−3	−2	4
−2	−1	1
−1	0	0
0	1	1
1	2	4

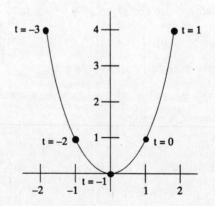

Solving for t in terms of x in the first parametric equation, we get

$t = x - 1$

Substituting for t in the second parametric equation we get

$y = (x - 1)^2 + 2(x - 1) + 1$

$y = x^2$

The graph of this equation is the same as the graph above.

TRIGONOMETRY

Trigonometric Identities

$\sin^2 \alpha + \cos^2 \alpha = 1$

$\tan \alpha = \dfrac{\sin \alpha}{\cos \alpha}$

$\cot \alpha = \dfrac{\cos \alpha}{\sin \alpha} = \dfrac{1}{\tan \alpha}$

$\csc \alpha = \dfrac{1}{\sin \alpha}$

$\sec \alpha = \dfrac{1}{\cos \alpha}$

$1 + \tan^2 \alpha = \sec^2 \alpha$

$1 + \cot^2 \alpha = \csc^2 \alpha$

One can find all the trigonometric functions of an acute angle when the value of any one of them is known.

For example, given a is an acute angle and $\csc \alpha = 2$, then

$\sin \alpha = \dfrac{1}{\csc \alpha} = \dfrac{1}{2}$

$\cos^2 \alpha + \sin^2 \alpha = 1, \cos \alpha = \sqrt{1 - \sin^2 \alpha} = \sqrt{1 - \left(\tfrac{1}{2}\right)^2} = \sqrt{1 - \tfrac{1}{4}} = \dfrac{\sqrt{3}}{2}$

$\tan \alpha = \dfrac{\sin \alpha}{\cos \alpha} = \dfrac{\tfrac{1}{2}}{\tfrac{\sqrt{3}}{2}} = \dfrac{1}{\sqrt{3}} = \dfrac{\sqrt{3}}{3}$

$\cot \alpha = \dfrac{1}{\tan \alpha} = \sqrt{3}$

$\sec \alpha = \dfrac{1}{\cos \alpha} = \dfrac{1}{\tfrac{\sqrt{3}}{2}} = \dfrac{2}{\sqrt{3}} = \dfrac{2\sqrt{3}}{3}$

For a given angle θ in standard position, the related angle of θ is the unique acute angle that the terminal side of θ makes with the x-axis.

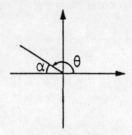

∠α is the related angle of ∠θ

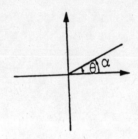

∠α is the related angle of ∠θ

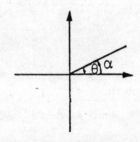

∠α is the related angle of ∠θ

Let θ be an angle in standard position and ø be the related angle of θ.

1. If θ is a first quadrant angle, then

    a) $\sin \theta = \sin \varnothing$

    b) $\cos \theta = \cos \varnothing$

    c) $\tan \theta = \tan \varnothing$

    d) $\cot \theta = \cot \varnothing$

    e) $\sec \theta = \sec \varnothing$

    f) $\csc \theta = \csc \varnothing$

2. If θ is a second quadrant angle, then

   a) $\sin \theta = \sin \phi$

   b) $\cos \theta = -\cos \phi$

   c) $\tan \theta = -\tan \phi$

   d) $\cot \theta = -\cot \phi$

   e) $\sec \theta = -\sec \phi$

   f) $\csc \theta = \csc \phi$

3. If θ is a third quadrant angle, then

   a) $\sin \theta = -\sin \phi$

   b) $\cos \theta = -\cos \phi$

   c) $\tan \theta = \tan \phi$

   d) $\cot \theta = \cot \phi$

   e) $\sec \theta = -\sec \phi$

   f) $\csc \theta = -\csc \phi$

4. If θ is a fourth quadrant angle, then

   a) $\sin \theta = -\sin \phi$

   b) $\cos \theta = \cos \phi$

   c) $\tan \theta = -\tan \phi$

   d) $\cot \theta = -\cot \phi$

   e) $\sec \theta = \sec \phi$

   f) $\csc \theta = -\csc \phi$

	sin	cos	tan	cot	sec	csc
$-\alpha$	$-\sin\alpha$	$+\cos\alpha$	$-\tan\alpha$	$-\cot\alpha$	$+\sec\alpha$	$-\csc\alpha$
$90°+\alpha$	$+\cos\alpha$	$-\sin\alpha$	$-\cot\alpha$	$-\tan\alpha$	$-\csc\alpha$	$+\sec\alpha$
$90°-\alpha$	$+\cos\alpha$	$+\sin\alpha$	$+\cot\alpha$	$+\tan\alpha$	$+\csc\alpha$	$+\sec\alpha$
$180°+\alpha$	$-\sin\alpha$	$-\cos\alpha$	$+\tan\alpha$	$+\cot\alpha$	$-\sec\alpha$	$-\csc\alpha$
$180°-\alpha$	$+\sin\alpha$	$-\cos\alpha$	$-\tan\alpha$	$-\cot\alpha$	$-\sec\alpha$	$+\csc\alpha$
$270°+\alpha$	$-\cos\alpha$	$+\sin\alpha$	$-\cot\alpha$	$-\tan\alpha$	$+\csc\alpha$	$-\sec\alpha$
$270°-\alpha$	$-\cos\alpha$	$-\sin\alpha$	$+\cot\alpha$	$+\tan\alpha$	$-\csc\alpha$	$-\sec\alpha$
$360°+\alpha$	$+\sin\alpha$	$+\cos\alpha$	$+\tan\alpha$	$+\cot\alpha$	$+\sec\alpha$	$+\csc\alpha$
$360°-\alpha$	$-\sin\alpha$	$+\cos\alpha$	$-\tan\alpha$	$-\cot\alpha$	$+\sec\alpha$	$-\csc\alpha$

Addition and Subtraction Formulas:

$$\sin(A \pm B) = \sin A \cos B \pm \cos A \sin B$$

$$\cos(A \pm B) = \cos A \cos B \mp \sin A \sin B$$

$$\tan(A \pm B) = \frac{\tan A \pm \tan B}{1 \mp \tan A \tan B}$$

$$\cot(A \pm B) = \frac{\cot A \cot B \mp 1}{\cot B \pm \cot A}$$

Double-Angle Formulas:

$$\sin 2A = 2 \sin A \cos A$$

$$\cos 2A = 2 \cos^2 A - 1 = 1 - 2 \sin^2 A = \cos^2 A - \sin^2 A$$

$$\tan 2A = \frac{2 \tan A}{1 - \tan^2 A}$$

Half-Angle Formulas:

$$\sin \frac{A}{2} = \pm \sqrt{\frac{1 - \cos A}{2}}$$

$$\cos \frac{A}{2} = \pm \sqrt{\frac{1 + \cos A}{2}}$$

$$\tan \frac{A}{2} = \pm \sqrt{\frac{1 - \cos A}{1 + \cos A}} = \frac{1 - \cos A}{\sin A} = \frac{\sin A}{1 + \cos A}$$

$$\cot \frac{A}{2} = \pm \sqrt{\frac{1 + \cos A}{1 - \cos A}} = \frac{1 + \cos A}{\sin A} = \frac{\sin A}{1 - \cos A}$$

Sum and Difference Formulas:

$$\sin \alpha + \sin \beta = 2 \sin\left(\frac{\alpha + \beta}{2}\right) \cos\left(\frac{\alpha - \beta}{2}\right)$$

$$\sin \alpha - \sin \beta = 2 \cos\left(\frac{\alpha + \beta}{2}\right) \sin\left(\frac{\alpha - \beta}{2}\right)$$

$$\cos \alpha + \cos \beta = 2 \cos\left(\frac{\alpha + \beta}{2}\right) \cos\left(\frac{\alpha - \beta}{2}\right)$$

$$\cos \alpha - \cos \beta = -2 \sin\left(\frac{\alpha + \beta}{2}\right) \sin\left(\frac{\alpha - \beta}{2}\right)$$

$$\tan \alpha + \tan \beta = \frac{\sin(\alpha + \beta)}{\cos \alpha \cos \beta}$$

$$\tan \alpha - \tan \beta = \frac{\sin(\alpha - \beta)}{\cos \alpha \cos \beta}$$

Product Formulas of Sines and Cosines:

$$\sin A \sin B = \frac{1}{2} \left[\cos(A - B) - \cos(A + B)\right]$$

$$\cos A \cos B = \frac{1}{2} \left[\cos(A + B) + \cos(A - B)\right]$$

$$\sin A \cos B = \frac{1}{2} \left[\sin(A + B) + \sin(A - B)\right]$$

$$\cos A \sin B = \frac{1}{2} \left[\sin(A + B) - \sin(A - B)\right]$$

Properties and Graphs of Trigonometric Functions

The *sine function* is the graph of $y = \sin x$. Other trigonometric functions are defined similarly.

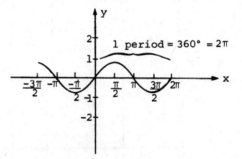

Sine Function

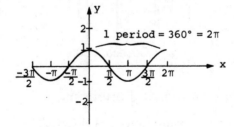

Cosine Function

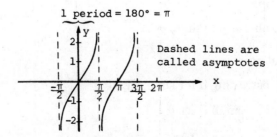

Tangent Function

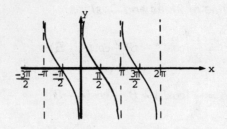

Cotangent Function

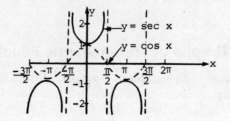

Secant Function

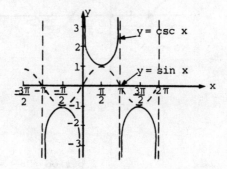

Cosecant Function

Composite Trigonometric Functions

Composite functions are functions of functions.

Example:

Evaluate $\cos (\sin \theta) = 1$

Solution:

Substituting α for $\sin \theta$, we get

$$\cos (\sin \theta) = \cos \alpha = 1$$

$$\alpha = 0, 2\pi$$

$$\sin \theta = 0 \text{ in rad.}$$

$$\theta = 0$$

$$\cos(\sin 0) = \cos(0) = 1$$

Composite trigonometric functions are used for calculus exercises. Composite functions are used to discuss inverse trigonometric functions below.

Inverse Trigonometric Functions

If $-1 < x < 1$, then there are infinitely many angles whose sine is x, as we can see by looking at the graph of the sine function.

$\arcsin x =$ the angle between $-\dfrac{\pi}{2}$ and $\dfrac{\pi}{2}$ whose sine is x.

$\text{arccsc } x =$ the angle between $-\dfrac{\pi}{2}$ and $\dfrac{\pi}{2}$ whose cosecant is x.

$\text{arccot } x =$ the angle between $-\dfrac{\pi}{2}$ and $\dfrac{\pi}{2}$ whose cotangent is x.

$\arccos x =$ the angle between 0 and π whose cosine is x.

$\text{arcsec } x =$ the angle between 0 and π whose secant is x.

$\arctan x =$ the angle between 0 and π whose tangent is x.

Example:

Evaluate $\arcsin \dfrac{1}{2}$.

Solution:

Since $\sin \dfrac{\pi}{6} = \dfrac{1}{2}$, $\arcsin \dfrac{1}{2} = \dfrac{\pi}{6}$.

The sine function and the arcsine function (abbreviated arcsin or $\sin^{-1}$) are inverses of each other in the sense that the composition of the two functions is the identity function (that is the function that takes x back to x).

$$\sin (\arcsin x) = x$$

$$\arcsin (\sin x) = x$$

Periodicity

The *period* of a (repeating) function, f, is the smallest positive number p such that $f(x) = f(x + p)$ for all x.

The period of the tangent and cotangent functions is π. This fact is clear from the graphs of the tangent and cotangent functions. Pick any angle, x, on the x-axis, and notice $x + \pi$ has the same tangent as x. The period of the other trigonometric functions is 2π.

If the period of a function f is p, and $g(x) = f(nx)$, then the period of g is p/n.

Example:

What is the period of $\sin 3x$?

Solution:

Since the period of $\sin x$ is 2π, the period of $\sin 3x$ is $\dfrac{2\pi}{3}$.

ELEMENTARY FUNCTIONS

Composition of Functions

Example:

Find $f(g(2))$ if

$$f(x) = x^2 - 3$$

$$g(x) = 3x + 1$$

Solution:

$$g(2) = 3(2) + 1 = 7$$

Substitute 7 for $g(2)$.

$$f(g(2)) = f(7) = 7^2 - 3 = 46$$

The functions f and g may be pictured as follows:

The composite function $f \bullet g$ may be pictured as follows:

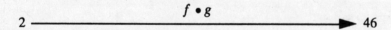

The composite function $f \bullet g$ is defined by

$$f \bullet g(x) = f(g(x))$$

Domain and Range

Suppose the numbers 2, 3, and 5 correspond with the numbers 1 and 2 under the function f according to the following diagram.

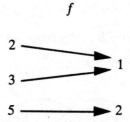

Then the domain of the function f is {2,3,5}, and the range of f is {1,2}. For each x in the domain of f, $f(x)$ is the number in the range of f that x corresponds to. So $f(2) = 1$.

Example:

Suppose the function f is defined by $f(x) = \sqrt{x-1}$. What is the largest possible domain and range of f if both x and $f(x)$ are required to be real numbers?

Solution:

$f(1) = 0$

So 1 corresponds with 0 under the function f. This fact may be visualized using the following diagram:

$$1 \xrightarrow{\quad f \quad} 0$$

The *domain* is the set of possible values of x. There is a real square root of $x - 1$ if and only if $x - 1$ is non-negative. So x must be greater than or equal to 1.

Domain = $\{x \mid x \geq 1\}$

The *range* is the set of all possible values of $f(x)$. We can arrange for the square root to be any non-negative number by choosing $x - 1$ to be the square of that number. So the range is all non-negative numbers.

Range = $\{f(x) \mid f(x) \geq 0\}$

MISCELLANEOUS TOPICS

Set Theory

The *union* of the sets A and B is the set of all points that are in A or in B.

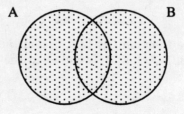

A ∪ B is shaded

Example:

What is the union of the sets {1,3,5} and {1,5,6}?

Solution:

{1,3,5,6}

The *intersection* of the sets A and B is the set of all points that are in both A and B.

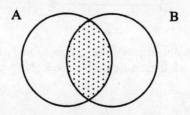

A ∩ B is shaded

Example:

What is the intersection of the sets {1,3,5} and {1,5,6}?

Solution:

{1,5}

Probability and Statistics

An *event* is a set of outcomes of an experiment.

The *probability* of an event is the fraction of the time the event will occur.

If there are finitely many equally likely outcomes of an experiment, then the probability of an event, E, is

$$P(E) = \frac{\text{number of outcomes in } E}{\text{total number of possible outcomes}}$$

Example:

A fair die is rolled. What is the probability the result is greater than 4?

Solution:

There are two outcomes greater than four, and there are six possible outcomes all together, (namely 1, 2, 3, 4, 5 and 6) so the probability is:

$P(\{5,6\}) = 2/6 = 1/3$

The probability that events A and B will both occur is the probability of $A \cap B$. That is,

$P(A \text{ and } B) = P(A \cap B)$

Example:

A fair die is rolled. What is the probability the result is greater than 4 and the result is even?

Solution:

The set of outcomes that are greater than four and even is $\{6\}$. So the probability is:

$P(\{6\}) = 1/6$

The probability that the event A or the event B will occur is:

$P(A \text{ or } B) = P(A \cup B)$

Example:

A fair die is rolled. What is the probability the result is greater than 4 or the result is even?

Solution:

The set of outcomes that are greater than 4 or even is $\{2,4,5,6\}$. So the probability is:

$$P(\{2,4,5,6\}) = \frac{4}{6} = \frac{2}{3}$$

The *conditional probability* of event B given event A is the fraction of the times that B occurs among just the times that A occurs.

The conditional probability of B given A is given by the formula:

$$P(B|A) = \frac{P(B \cap A)}{P(A)}$$

Example:

A fair die is rolled. What is the conditional probability the result is even given the result is greater than 3?

Solution:

The outcomes greater than 3 are $\{4,5,6\}$. Among those outcomes, the outcome is even $\frac{2}{3}$ of the time.

$$P(B|A) = \frac{P(B \cap A)}{P(A)}$$

Suppose we roll two dice 36 times and get the following sums:

2,
3, 3, 3,
4,
5, 5, 5, 5, 5,
6, 6, 6, 6,
7, 7, 7, 7, 7, 7,
8, 8, 8, 8, 8,
9, 9, 9, 9, 9, 9,
10, 10,
11, 11,
12

The *mean* of this data is the sum of the outcomes divided by the number of outcomes (that is, divided by 36). The mean is approximately 7.1.

The *mode* of this data is the most common outcome. Since 7 and 9 are equally common the data has two modes. Such data is sometimes called *bi-modal*.

The *median* of this data is the outcome for which half the outcomes are less than or equal to that score, and half the outcomes are greater than or equal to that score. The median is 7.

Logic and Proof

In mathematics, if two statements are connected with "or," then the compound statement is true if either or both of the original statements is true.

Examples:

> 2 < 3 or 5 < 7   True
>
> 2 < 3 or 5 < 3   True
>
> 2 < 1 or 5 < 7   True
>
> 2 < 1 or 5 < 3   False

This mathematical tradition is sometimes followed in common speech. Suppose people are allowed to attend a movie if they are over 17 or accompanied by their parent, and an 18-year-old goes to the theater alone. Then the statement that the person is over 17 is true, and the statement that the person is accompanied by his or her parent is false.

The statement that the person is over 17 or accompanied by a parent is true. This example may help you remember how to handle statements containing the word "or."

If two statements are connected by "and," then the compound statement is true if and only if both statements are true.

Examples:

> 2 < 3 and 5 < 7   True
>
> 2 < 3 and 5 < 3   False
>
> 2 < 1 and 5 < 7   False
>
> 2 < 1 and 5 < 3   False

DeMorgan's Laws:

The compound statement "p and q" is false if and only if "not p" or "not q" is true.

The compound statement "p or q" is false if and only if "not p" and "not q" are true.

The statement "if p is true, then q is true" is a true statement whenever q is true, and whenever p is false. For example, the statement "If you are an elephant then I will eat my hat" is true, even though I will not eat my hat. The statement only claims I will eat my hat when you are an elephant, and there are no such times.

If we can show the statement "if q is false, then p is false" is a true statement, then we may conclude the statement "if p is true, then q is true" is also a true statement. (If p is true, then the first statement doesn't permit q to be false.)

This principle is used in indirect proofs.

Sequences and Limits

A list of numbers is called a *sequence* or progression. The list may stop or it may be infinitely long. An *arithmetic sequence* is a sequence for which the difference between terms and their successors is constant.

The following sequence is arithmetic, because the difference between successive terms is always two.

3, 5, 7, 9, 11

Suppose we have an arithmetic sequence where

a is the first term

d is the common difference, and

k is the number of terms

Then the sequence is

$$a, a + d, a + 2d, \dots , a + (k-1)d$$

To get the second term, we don't add d to a twice; we add it only once. To get the last term, we don't add d to a k times; we add it only $k - 1$ times. So the last term is

last term $= a + (k - 1)d$

The mean of the first and last terms is

$$\frac{a + [a + (k - 1)d]}{2}$$

The second term is d larger than the first, and the next to last term is d smaller than the last, so the mean of the second term and the next to last term is the same as the mean of the first and last terms. Continuing this reasoning, you may see that the average size term for the whole sequence is the same as the mean of the first and last.

Since there are k terms in the arithmetic sequence, the sum is k times the mean of the sequence.

$$\text{sum of } k \text{ terms} = k \, \frac{a + [a + (k - 1)d]}{2}$$

A *geometric sequence* is a sequence for which the ratio of any term with its successor is a constant.

3, 6, 12, 24, 48

Suppose we have a geometric sequence where

a is the first term

r is the common ratio

k is the number of terms

Then the sequence is

$$a, ar, ar^2, ar^3, \ldots ar^{k-1}$$

Then the last term is

$$ar^{k-1}$$

and the sum of first n terms is

$$\frac{a(r^n - 1)}{r - 1}; \text{ if } |r| > 1$$

or $\quad \dfrac{a(1 - r^n)}{1 - r}; \text{ if } |r| < 1$

If $|r| < 1$, and the geometric sequence continues to infinitely many terms, then the sum of all the terms of this infinite geometric sequence is

$$\frac{a}{1 - r}; \text{ if } |r| < 1$$

The sequence

$$\frac{1}{2}, \frac{1}{3}, \frac{1}{4}, \frac{1}{5}, \frac{1}{6}, \ldots$$

is an *infinite sequence*. The three dots at the end indicate it goes on forever. Terms of the sequence get closer and closer to zero, so zero is said to be the *limit* of the sequence.

Definition:

The limit of an infinite sequence is the number (if there is one) that the sequence gets closer and closer to.

More precise definition:

The limit of an infinite sequence is L if, no matter how close we want the terms to be to L, they will be that close from some term on.

Even more precise definition:

The limit of an infinite sequence is L if for each positive distance d, there exists a positive integer n such that the distance from each term after the nth term to L is less than d.

Elementary Number Theory

An integer greater than 1 that has no factors other than 1 and itself is called *prime*.

The prime numbers in the following list are underlined.

2 3 4 5 6 7 8 9 10 11 12 13 14 15 16 17 18 19 20 21 22 23 24 25

Definition:

An integer greater than 1 that is not prime is called *composite*.

The numbers in the list above that are not underlined are composite.

Definition:

If a and b have no prime factors in common, then a and b are *relatively prime*.

Example:

Eight and nine are relatively prime, because

8 = 2(2)(2), and

9 = 3(3).

There are no primes in the first factorization that appear in the second factorization.

The fraction a/b is in lowest terms if and only if a and b are relatively prime. So the fraction $8/9$ is in lowest terms.

THE SAT SUBJECT TEST IN

Math
Level 2

PRACTICE TEST 1

This test is also on CD-ROM in our special interactive SAT Math Level 2 TEST*ware*®. It is highly recommended that you first take this exam on computer. You will then have the additional study features and benefits of enforced timed conditions and instant, accurate scoring. See page 2 for guidance on how to get the most out of our SAT Math Level 2 software.

SAT Mathematics Level 2

Practice Test 1

Time: 1 Hour
50 Questions

DIRECTIONS: Choose the best answer for each question and mark the letter of your selection on the corresponding answer sheet in the back of the book.

NOTES:

(1) Some questions require the use of a calculator. You must decide when the use of your calculator will be helpful.

(2) You may need to decide which mode your calculator should be in—radian or degree.

(3) All figures are drawn to scale and lie in a plane unless otherwise stated.

(4) The domain of any function f is the set of all real numbers x for which $f(x)$ is a real number, unless other information is provided.

REFERENCE INFORMATION: The following information may be helpful in answering some of the questions.

Volume of a right circular cone with radius r and height h	$V = \dfrac{1}{3}\pi r^2 h$
Lateral area of a right circular cone with circumference of the c and slant height l	$S = \dfrac{1}{2}cl$
Volume of a sphere with radius r	$V = \dfrac{4}{3}\pi r^3$
Surface Area of a sphere with r	$S = 4\pi r^2$
Volume of a pyramid with base area B and height h	$V = \dfrac{1}{3}Bh$

1. Find the center of the ellipse given by the equation

    $x^2 + x + 3y + 2y^2 - 1 = 0.$

    (A) $\left(\dfrac{1}{2}, -\dfrac{3}{2}\right)$ (D) $\left(-\dfrac{1}{2}, -\dfrac{3}{4}\right)$

    (B) $\left(\dfrac{1}{4}, \dfrac{3}{2}\right)$ (E) $\left(1, -\dfrac{3}{4}\right)$

    (C) $\left(\dfrac{1}{2}, -\dfrac{3}{4}\right)$

2. One plane flies at a ground speed 75 miles per hour faster than another. On a particular flight, the faster plane requires 3 hours and the slower one 3 hours and 36 minutes. What is the distance of the flight?

    (A)   450 miles   (D)   1,450 miles

    (B)   1,350 miles   (E)   375 miles

    (C)   1,000 miles

3. Which value of x fits the equation below for $0 \le x \le 90°$?

    $2 \sin x + 2 \cos x = 1 + \sqrt{3}$

    (A)   30°   (D)   0°

    (B)   45°   (E)   90°

    (C) $-90°$

4. A telephone number consists of 7 digits. How many different telephone numbers exist if each digit appears only one time in the number?

    (A)   7!   (D) 7^7

    (B) 10^7 (E) $\dfrac{10!}{3!}$

    (C)   70

5. If "Δ" is an operator that transforms a, b according to the rule $a \, \Delta \, b = a - b$, determine $(a \, \Delta \, b) \, \Delta \, ((a \, \Delta \, a) \, \Delta \, b)$.

(A) $a - 2b$ (D) $2a$

(B) $2b$ (E) a

(C) $2a - b$

6. Find the area of the circle

$$x^2 + 6x + y^2 + 8y + 6 = 0.$$

(A)   16   (D)   50

(B)   25   (E)   60

(C)   42

7. What is $\dfrac{15! \times 6!}{13!}$?

(A)   56,370   (D)   1,456,782

(B)   6,537,380   (E)   151,200

(C)   89,076,550

8. The solution set for x in the inequality $| \, x^2 - 3 \, | < 1$ is

(A) $\{x \mid -\sqrt{2} < x < \sqrt{2}\}$.

(B) $\{x \mid -2 < x < -\sqrt{2}\}$.

(C) $\{x \mid -2 < x < -\sqrt{2} \ \text{or} \ \sqrt{2} < x < 2\}$.

(D) $\{x \mid -\sqrt{2} < x < \sqrt{2} \ \text{or} \ -2 < x < 2\}$.

(E) $\{x \mid \sqrt{2} < x < 2\}$.

9. If a right triangle is rotated 360° around one of its sides the solid generated is a

(A) pyramid.

(D) sphere.

(B) parallelepiped.

(E) cylinder.

(C) cone.

10. What is the value of k if the line $y = 2x + k$ has to be tangent to the hyperbola $x^2 - y^2 = 1$?

(A) $\sqrt{3}$

(D) $2\sqrt{3}$

(B) $-\sqrt{3}$

(E) $-2\sqrt{3}$

(C) $\pm\sqrt{3}$

11. If $f(x) = \ln x + 3$, find $\log(f^{-1}(10))$.

(A) 2.35

(D) 8.13

(B) 3.04

(E) 15.2

(C) 5.24

12. If $\log_8 3 = x\log_2 3$ then x equals

(A) 4

(D) $\log_8 9$

(B) $\log_4 3$

(E) $\frac{1}{3}$

(C) 3

13. $\log_3(81)^{-2.3} =$

(A) 8.0

(D) -9.2

(B) 3.0

(E) -4.5

(C) 5.6

14. $\log_2 \dfrac{\sqrt{2}}{8} =$

(A) $\dfrac{5}{2}$

(D) $-\dfrac{5}{2}$

(B) $\dfrac{1}{2}$

(E) $-\dfrac{1}{2}$

(C) $\dfrac{3}{2}$

15. Jack is 5 times as old as Bill. Ten years from now, Jack will be at least 3 times as old as Bill will be then. At least how old is Jack now?

(A) 10 years

(D) 20 years

(B) 50 years

(E) 25 years

(C) 30 years

16. $\sin(x + y) = 0.9659$, $\sin(x) = 0.5$. Find $\cos(y)$.

(A) 0.425

(D) 0.816

(B) 1

(E) 0.5

(C) 0.707

17. The graph of the parabola $y^2 = -4x$ could be represented by which of the graphs below?

(A)

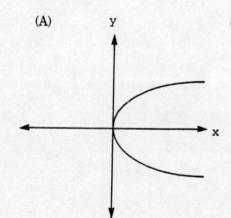

(B)

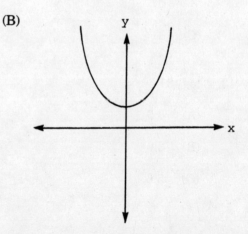

(C)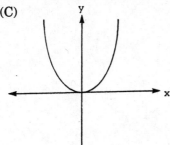

(D)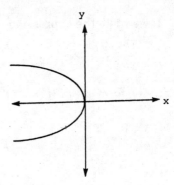

(E)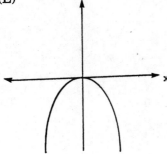

18. Which of the following sentences below concerning the equation of an ellipse

$$\frac{x^2}{a^2} + \frac{y^2}{b^2} = 1,$$

is true when $a = b = r$?

I.   The equation reduces to $x^2 + y^2 = r^2$.

II. The equation reduces to an equation of a circle with center at $(0, 0)$ and radius r.

III. The equation reduces to an equation of a circle, center $(0, r)$.

(A) I only.

(B) I and II only.

(C) II and III only.

(D) III only.

(E) All of the statements are incorrect.

19. If we write the expression

$$\frac{1}{1-i} - \frac{1}{i}$$

in the form $a + bi$ ($i^2 = -1$) the result will be

(A) $3 + 2i$

(D) $\frac{1}{2} + \frac{3}{2}i$

(B) $\frac{5}{3} - \frac{1}{3}i$

(E) $\frac{1}{2} + 2i$

(C) $\frac{1}{2} + 5i$

20. Calculate $\log 5^{-3} + \ln \frac{2}{e^{-3}} + 4^{-0.3}$.

(A)   2.256

(D)   8.0

(B)   3.26

(E)   100.2

(C)   7.293

21. Which function(s) below is(are) symmetric with respect to the origin?

I. $f(x) = x^3 - x$

II. $f(x) = 2x + x^5$

III. $f(x) = 2x + 4$

(A)   I and II.

(D)   II and III.

(B)   I only.

(E)   I, II, and III.

(C)   I and III.

22. A single fair die is tossed. Find the probability of obtaining either a 3 or a 5.

(A) $\frac{1}{3}$

(D) $\frac{2}{5}$

(B) $\frac{2}{3}$

(E) $\frac{1}{6}$

(C) $\frac{1}{36}$

23. The closest approximation of $(1.001)^8$ is

   (A) 1.006
   (B) 1.007
   (C) 1.008

   (D) 1.009
   (E) 1.010

24. $\sin\left(\dfrac{1}{2}\pi + t\right) =$

   (A) $\sin\dfrac{\pi}{2} + \sin t$
   (B) $\sin 2t$
   (C) $\sin t$

   (D) $\cos 2t$
   (E) $\cos t$

25. A graph can be described by

   $x = \ln(t) + 1,\ y = \log(t) - 1.$

   What is x when $y = 0.5$?

   (A) 4.45
   (B) 12.2
   (C) 0.25

   (D) 8.0
   (E) 1.0

26. The solution of $\dfrac{2x+1}{x^2-4} > 0$ is

   (A) $x > 2$
   (B) $x > 1$
   (C) $x > 2$ or $x < \dfrac{1}{2}$

   (D) $x > 2$ or $-2 < x < -\dfrac{1}{2}$
   (E) $x > 2$ or $x < -2$

27. A rectangle has dimensions 3, 4, and 5. Find its diagonal.

   (A) 4.33
   (B) 9.41
   (C) 3.14

   (D) 5.25
   (E) 7.07

28. What is the slope of the line

$$15x + 37y - 23 = 0?$$

(A)   0.405

(B) -0.405

(C)   2.467

(D)   15

(E)   37

29. If $x - 2$ is a factor of

$$x^3 - 7x^2 + kx - 12,$$

then k is

(A)   32

(B)   16

(C)   2

(D) -32

(E) -16

30. Which of the graphs of the functions below has symmetry with respect to the y-axis?

(A) $y = x + 2$

(B) $y = 2^x$

(C) $y = x^2 + x$

(D) $y = x^2$

(E) $y = \sin x$

31. If the probability of a certain team winning is $3/4$, what is the probability that this team will win its first 3 games and lose the fourth?

(A) $\dfrac{3}{256}$

(B) $\dfrac{9}{256}$

(C) $\dfrac{1}{256}$

(D) $\dfrac{81}{256}$

(E) $\dfrac{27}{256}$

32. Which of the expressions below is equivalent to $\csc^2 x$?

(A) $-\tan^2 x + \sec x$

(B) $1 - \cot^2 x$

(C) $1 + \cot^2 x$

(D) $-\dfrac{1}{\sin^2 x}$

(E) $-\dfrac{1}{\cos^2 x}$

33.   Let $f(x) = \dfrac{\sin(x)\cos(2x)}{\tan(3x)}$. Find $f(a)$, $a = 2.5$ in radian measure.

(A)   0.89

(D)   0.522

(B)   0.063

(E)   0.12

(C)   1.235

34.   The distance of the plane

$$2x - 3y + 5z + 3 = 0$$

from the point $(2, 3, -1)$ is

(A)   1.136

(D)   8.0

(B)   2.545

(E)   12.45

(C)   3.65

35.   What is the equation of the line which is parallel to $6x + 3y = 4$, and has a y-intercept of -6?

(A) $y = -2x + \dfrac{4}{3}$

(D) $y = -2x - 6$

(B) $y = 2x + \dfrac{4}{3}$

(E) $y = -2x + 6$

(C) $y = -2x - \dfrac{4}{3}$

36.   Find the value of the function $f(x) = x^{\log x - 3}$ when $x = 2$.

(A)   0.535

(D)   2.78

(B)   0.756

(E)   5.54

(C)   1.534

37. Find a, b for the parabola

$$y = ax^2 + bx + 3$$

if the vertex is (2, 4).

(A) $a = -\dfrac{1}{4}, b = 2$

(D) $a = -\dfrac{1}{3}, b = 1$

(B) $a = -1, b = -2$

(E) $a = -\dfrac{1}{4}, b = 1$

(C) $a = 1, b = 2$

38. The left side of the equation

$$x^4 + ax^3 + 2x^2 + 6x + 8 = 0$$

has a factor $(x - 1.5)$. Find a.

(A) -8.9

(D) 4.2

(B) 2.3

(E) -7.87

(C) 6.6

39. If S_{AP} is the sum of the first four terms of an arithmetic progression and S_{GP} is the sum of the first four terms of a geometric progression and

$$\frac{S_{AP}}{S_{GP}} = \frac{2}{3},$$

considering $r = 2$, for both of them, determine $S_{AP} + S_{GP}$.

(A) 30

(D) 50

(B) 20

(E) 60

(C) 40

40. $\dfrac{\sqrt[3]{-16}}{\sqrt[3]{-2}} =$

(A) 2

(D) $\sqrt[3]{2}$

(B) -2

(E) $\sqrt[3]{-2}$

(C) $2\sqrt{3}$

41. If the functions f and g are defined by

$$f(x, y) = x^2 + 2y^2 - y \text{ and } g(h) = h + 7,$$

what is $f(3, g(-2))$?

(A) 24   (D) 64

(B) 54   (E) 19

(C) 59

42. Find $\left\{ x \left| \dfrac{2}{x+1} - 3 = \dfrac{4x+6}{x+1} \right. \right\}$.

(A) $x = -2$ (D) $x = 7$

(B) ϕ (E) $x = -7$

(C) $x = 1$

43. A circular region rotated 360° around its diameter as an axis generates a

(A) cube.

(B) rectangular parallelepiped.

(C) cone.

(D) sphere.

(E) cylinder.

44. If $f(x) = 2x + 4$ and $g(x) = x^2 - 2$, then $(f \circ g)(x)$, where $(f \circ g)(x)$ is a composition of functions, is

(A) $2x^2 - 8$ (D) $2x^3 + 4x^2 - 4x - 8$

(B) $2x^2 + 8$ (E) $4x^2 + 16x + 14$

(C) $2x^2$

45. Use the diagrams below to find the value of $\cos \dfrac{1}{12}\pi$.

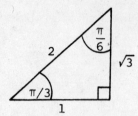 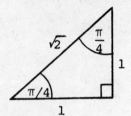

(A) $\dfrac{\sqrt{2}}{4}\left(1+\sqrt{3}\right)$

(D) $\dfrac{\sqrt{3}}{\sqrt{2}}$

(B) $\dfrac{1}{2}\left(1+\sqrt{3}\right)$

(E) $\dfrac{1}{2}\left(1-\sqrt{3}\right)$

(C) $1+\sqrt{3}$

46. Given $\log_{10}2 = 0.3010$, find $\log_{10}32$.

(A) 0.6020

(D) 30.10

(B) 3.010

(E) 0.03010

(C) 1.5050

47. Find the value of an acute angle x when $2 \cos x - \sqrt{2} = 0$.

(A) 30°

(D) 25°

(B) 60°

(E) 27.5°

(C) 45°

48. $\cos\left(\dfrac{\pi}{2} - \theta\right) =$

(A) $\sin \theta$

(D) $\sin\left(\dfrac{\pi}{2} - \theta\right)$

(B) $\cos \theta$

(E) $\sin 2\theta$

(C) $\sin \theta \cos \theta$

49. If $f(x) = x - 2x^2 + 2$, then $f(a - 2) =$

    (A) $(a-2)^2 + 2$ (D) $3a^2 - 9a + 8$

    (B) $9a - 2a^2 - 8$ (E) $-9a + 2a^2 - 8$

    (C) $2a^2 + 10a - 8$

50. $f = \{(x, y) : y = x^2$ and $x \geq 0\}$. Find f^{-1}.

    (A) $\{(x, y): y = +\sqrt{x}\}$ (D) $\{(x, y) : y \geq \sqrt{x}\}$

    (B) $\{(x, y) : y = \pm\sqrt{x}\}$ (E)   None of the above.

    (C) $\{(x, y) : y = -\sqrt{x}\}$

SAT Mathematics Level 2

Practice Test 1
ANSWER KEY

1.	(D)	14.	(D)	27.	(E)	40.	(A)
2.	(B)	15.	(B)	28.	(B)	41.	(B)
3.	(A)	16.	(C)	29.	(B)	42.	(B)
4.	(E)	17.	(D)	30.	(D)	43.	(D)
5.	(E)	18.	(B)	31.	(E)	44.	(C)
6.	(E)	19.	(D)	32.	(C)	45.	(A)
7.	(E)	20.	(A)	33.	(B)	46.	(C)
8.	(C)	21.	(A)	34.	(A)	47.	(C)
9.	(C)	22.	(A)	35.	(D)	48.	(A)
10.	(C)	23.	(C)	36.	(A)	49.	(B)
11.	(B)	24.	(E)	37.	(E)	50.	(A)
12.	(E)	25.	(A)	38.	(E)		
13.	(D)	26.	(D)	39.	(D)		

DETAILED EXPLANATIONS
OF ANSWERS

1. **(D)**

Rewrite the equation in standard form and complete the squares:

$$(x^2 + x + \frac{1}{4}) - \frac{1}{4} + 2(y^2 + \frac{3}{2}y) - 1 = 0$$

$$(x + \frac{1}{2})^2 - \frac{1}{4} + 2(y^2 + \frac{3}{2}y + \frac{9}{16} - \frac{9}{16}) - 1 = 0$$

$$(x + \frac{1}{2})^2 + 2\left[(y + \frac{3}{4})^2 - \frac{9}{16}\right] - \frac{5}{4} = 0$$

$$(x + \frac{1}{2})^2 + 2(y + \frac{3}{4})^2 - \frac{9}{8} - \frac{5}{4} = 0$$

$$\frac{(x + \frac{1}{2})^2}{2} + (y + \frac{3}{4})^2 - \frac{19}{16} = 0$$

$$\frac{(x - (-\frac{1}{2}))^2}{\frac{19}{8}} + \frac{(y - (-\frac{3}{4}))^2}{\frac{19}{16}} = 1$$

The standard form of an ellipse is

$$\frac{(x - h)^2}{a^2} + \frac{(y - k)^2}{b^2} = 1,$$

where (h, k) is the center. Hence, the center is

$$\left(-\frac{1}{2}, -\frac{3}{4}\right).$$

2. **(B)**

If the velocity of the fast plane is v_1 and that of the slow plane is v_2, we have

$$v_1 = 75 + v_2 \tag{1}$$

since the fast plane is 75 miles per hour faster. The distance is the same for each plane. The distance is $d = v \times t$, where v is the velocity and t is the time. Note that the time for the slower plane expressed in hours is

$$3\frac{36}{60} = 3\frac{3}{5} \text{ hours.}$$

Hence,

$$3v_1 = 3\frac{3}{5}v_2 \tag{2}$$

Equations (1) and (2) constitute a system of two equations in two variables. Solving equation (2) for v_1, we have

$$v_1 = \frac{6}{5}v_2$$

Substituting this value in equation (1), we have

$$\frac{6}{5}v_2 = 75 + v_2$$

$$\frac{1}{5}v_2 = 75$$

$$v_2 = 375$$

$$v_1 = 75 + 375 = 450$$

Hence, the length of the trip is $3 \times 450 = 1{,}350$ miles.

Check: The fast plane, velocity 450 miles per hour, is 75 miles per hour faster than the slow one, velocity 375 miles per hour. In 3 hours the fast plane travels $3 \times 450 = 1{,}350$ miles. In 3 hours 36 minutes the slow plane travels the same distance; that is, $3^3/_5 \times 375 = 1{,}350$ miles.

3.   **(A)**
Since $2\sin x + 2\cos x = 1 + \sqrt{3}$, we can divide both sides by 2 and obtain:

$$\sin x + \cos x = \frac{1+\sqrt{3}}{2}$$

By squaring both sides:

$$(\sin x + \cos x)^2 = \left(\frac{1+\sqrt{3}}{2}\right)^2$$

$$\sin^2 x + 2\sin x \cos x + \cos^2 x = \frac{1 + 2\sqrt{3} + 3}{4}$$

$$= \frac{4 + 2\sqrt{3}}{4} = 1 + \frac{\sqrt{3}}{2}$$

Using the trigonometric property $\sin^2 x + \cos^2 x = 1$

$$2\sin x \cos x + 1 = 1 + \frac{\sqrt{3}}{2}$$

and if we remember that $2\sin x \cos x = \sin 2x$

$$2\sin x \cos x = \sin 2x = \frac{\sqrt{3}}{2}$$

Since $2x$ has to be in the first quadrant, $2x = 60°, x = 30°$.

4.   **(E)**
There are 10 integers from 0 to 9, hence there are 10 choices for the first digit of the phone number. Since no digit can be repeated we have only 9 choices for the second digit. Similarly, there are 8 choices for the third digit and so on. Since there are 7 digits in a phone number we have

$$10 \times 9 \times 8 \times 7 \times 6 \times 5 \times 4 = \frac{10!}{3!}$$

possible phone numbers.

5.   **(E)**
According to the rule given, and solving within the parentheses first:

$(a \, \Delta \, b) \, \Delta \, ((a \, \Delta \, a) \, \Delta \, b)$

$(a - b) \, \Delta \, (0 \, \Delta \, b)$

$(a - b) \, \Delta \, (- b)$

$(a - b) + b = a - b + b = a$

6.   **(E)**
The equation of the circle can be written as

$$x^2 + 6x + 3^2 + y^2 + 8y + 4^2 + 6 - 9 - 16 = 0$$

or

$$(x + 3)^2 + (y + 4)^2 = 19$$

Thus, the radius of the circle is $r = \sqrt{19}$ and the area is

$$\pi \times r^2 = 3.1416 \times 19 = 59.69.$$

So, the correct choice is (E).

7.   **(E)**

$$\frac{15! \times 6!}{13!} = 15 \times 14 \times 6 \times 5 \times 4 \times 3 \times 2 \times 1 = 151{,}200$$

8.   **(C)**
    If $|\ x^2 - 3\ | < 1$ we have two possibilities:

$$|\ x^2 - 3\ | = (x^2 - 3) \text{ or } -(x^2 - 3)$$

   a) $x^2 - 3 < 1$

   $x^2 - 4 < 0$

   $-2 < x < 2$

   b) $-(x^2 - 3) < 1$

   $-x^2 + 3 < 1$

   $-x^2 + 2 < 0$

   $x^2 - 2 > 0$

   $x < -\sqrt{2} \text{ or } x > \sqrt{2}$

The solution can be represented graphically as shown below:

9.   **(C)**
    The figure below shows a right triangle being rotated, as indicated by the arrow.

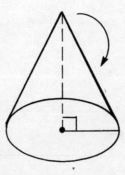

The figure generated is a cone.

10. **(C)**

The line $y = 2x + k$ will be tangent to the hyperbola if the system below has only one solution.

$$\begin{cases} x^2 - y^2 = 1 \\ y = 2x + k \end{cases}$$

By substitution we obtain:

$$x^2 - (2x + k)^2 = 1$$

$$3x^2 + 4kx + (k^2 + 1) = 0$$

Since this has to have only one solution $\Delta = 0$:

$$\Delta = (4k)^2 - 4 \times 3 \times (k^2 + 1)$$

$$= 4k^2 - 12$$

$$= 0$$

$$k = \pm \sqrt{3}$$

11. **(B)**

You need to solve this problem in two steps. First, find $f^{-1}(10)$. This can be done quite easily because

$$10 = \ln x + 3$$

Will give you directly

$$x = \ln^{-1}(10 - 3) = 1096.6.$$

The second step is to put $f^{-1}(10) = 1096.6$ into the log function. So, we have

$$\log(f^{-1}(10)) = \log(1096.6) = 3.04.$$

12. **(E)**

Let $y = \log_8 3 = x \log_2 3$.

Then $8^y = 3 \Rightarrow 2^{3y} = 3$ (1)

and $y = x\log_2 3 \Rightarrow 2^y = 3^x$ (2)

Substituting the expression for 2^y in (2) into (1) we obtain

$$3 = (2^y)^3 = (3^x)^3 = 3^{3x}.$$

Hence $3x = 1 \Rightarrow x = \dfrac{1}{3}.$

$$\log a^b = b \cdot \log a$$

---

13. **(D)**

Simplify:

$$\log_3(81)^{-2.3} = -2.3\log_3 3^4$$
$$= -2.3 \times 4\log_3 3$$
$$= -2.3 \times 4$$
$$= -9.2$$

14. **(D)**

The expression

$$\log_2 \frac{\sqrt{2}}{8}$$

can be rewritten as:

$$\log_2 \frac{2^{\frac{1}{2}}}{2^3}$$

and by subtracting the exponents:

$$\log_2 2^{\frac{1}{2}-3} = \log_2 2^{-\frac{5}{2}} = \frac{-5}{2}$$

15. **(B)**

Let J and B represent Jack's and Bill's ages respectively. We know that Jack is 5 times as old as Bill and can therefore conclude:

$$J = 5B.$$

We also know that in 10 years Jack will be $J + 10$ years old and Bill will be $B + 10$ years old. At that time Jack will be at least 3 times as old as Bill. So

$$(J + 10) \geq 3(B + 10).$$

We know $J = 5B$ and may therefore substitute this value into the second equation to obtain

$$((5B) + 10) \geq 3B + 30$$
$$2B \geq 20$$
$$B \geq 10$$

Bill is at least 10 years old now, so Jack must be at least 50 years old.

16.   **(C)**

$x = \sin^{-1}(0.5) = 30°.$

$x + y = \sin^{-1}(0.9659) = 75°.$

Then find y to be

$y = 75° - 30° = 45°.$

Then $\cos(y) = \cos(45°) = 0.707.$

17.   **(D)**

The point $(0, 0)$ satisfies the equation $y^2 = -4x$, so the graph has to pass through the origin, which excludes alternative (B). Since the graph is of the form $y^2 = ax$, the parabola should have concavity to the right or to the left, but the negative sign indicates that such graph should exist only for negative values of x. Therefore, the concavity is to the left as shown below.

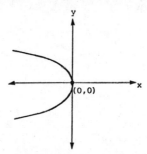

18.   **(B)**

If $a = b = r$, the given equation becomes

$$\frac{x^2}{a^2} + \frac{y^2}{b^2} = \frac{x^2}{r^2} + \frac{y^2}{r^2} = 1. \tag{1}$$

Multiplying the last branch of the equality in equation (1) by r^2 yields

$$x^2 + y^2 = r^2. \tag{2}$$

This is the equation of a circle with center at $C(0, 0)$ and radius r. Hence, we see that a circle is a special case of an ellipse.

19.   **(D)**

If we multiply the first term by

$$\frac{(1+i)}{(1+i)}$$

and the second by

$$\frac{i}{i}$$

we will obtain:

$$\frac{1}{1-i} \times \frac{1+i}{1+i} - \frac{1}{i} \times \frac{i}{i} = \frac{1+i}{1-(-1)} + \frac{i}{1}$$

$$= \frac{1+i}{2} + \frac{i}{1}$$

$$= \frac{1+i+2i}{2}$$

$$= \frac{1+3i}{2}$$

$$= \frac{1}{2} + \frac{3}{2}i$$

20.   **(A)**

The first term:

$$\log 5^{-3} = -3\log 5 = -2.097$$

The second term:

$$\ln \frac{2}{e^{-3}} = \ln 2 + 3 = 0.693 + 3 = 3.693$$

The last term:

$$4^{-0.3} = \frac{1}{4^{0.3}} = \frac{1}{1.5157} = 0.6598$$

Finally, the total:

$$-2.097 + 3.693 + 0.6598 = 2.256$$

21.   **(A)**

A function is symmetric with respect to the origin if replacing x by $-x$ and y by $-y$ produces an equivalent function.

(I) $y = f(x) = x^3 - x$

$(-y) = (-x)^3 - (-x)$

$-y = -x^3 + x$

$y = x^3 - x$

(I) is symmetric.

(II) $y = f(x) = 2x + x^5$

$-y = 2(-x) + (-x)^5$

$$-y = -2x - x^5$$

$$y = 2x + x^5$$

(II) is symmetric.

(III) $\qquad y = f(x) = 2x + 4$

$$-y = 2(-x) + 4$$

$$-y = -2x + 4$$

$$y = 2x - 4$$

(III) is not symmetric.

22. **(A)**

The die may land in any of 6 ways for a single toss. The probability of obtaining a 3 is:

$$P(3) = \frac{\text{Number of ways of obtaining a 3}}{\text{Number of ways the die may land}}$$

$\therefore \qquad P(3) = \dfrac{1}{6}$, similarly $P(5) = \dfrac{1}{6}$

The probability therefore of obtaining either a 3 or a 5 is

$$P(3) + P(5),$$

which equals

$$\frac{1}{6} + \frac{1}{6} = \frac{2}{6} = \frac{1}{3}.$$

23. **(C)**

$$(1.001)^8 = (1 + 0.001)^8$$

$$\approx 1^8 + 8(1)^7 \,(0.001) + \binom{8}{2} (1)^6 \,(0.001)^2$$

$$\approx 1 + 0.008$$

$$= 1.008$$

24. **(E)**

$$\sin\left(\frac{\pi}{2} + t\right) = \sin\frac{\pi}{2}\cos t + \cos\frac{\pi}{2}\sin t$$

But $\quad \sin\dfrac{\pi}{2} = 1$ and $\cos\dfrac{\pi}{2} = 0$

$\therefore \quad \sin\left(\dfrac{\pi}{2} + t\right) = \cos t.$

25.　(A)

Here x and y are both functions of a single variable t. They are related to each other through t. Thus, given one (either x or y), you can always find the other. Here, $y = 0.5$ is given. We can find

$$t = \log^{-1}(y + 1) = \log^{-1}(1.5) = 31.6.$$

Then, we find

$$x = \ln(31.6) + 1 = 4.4538.$$

Rounding this number, we get 4.45.

26.　(D)

If

$$\dfrac{2x + 1}{x^2 - 4} > 0$$

then either

$$2x + 1 > 0 \text{ and } x^2 - 4 > 0$$

or $\quad 2x + 1 < 0$ and $x^2 - 4 < 0.$

First case:

$$2x + 1 > 0 \text{ and } x^2 - 4 > 0$$

$$2x + 1 > 0$$

$$2x > -1$$

$$x > -\dfrac{1}{2} \qquad\qquad \text{(I)}$$

The equation $x^2 - 4 > 0$ can be represented as shown in the figure:

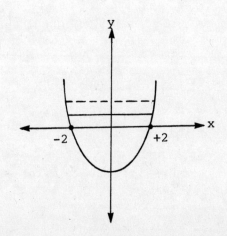

If $x < -2$ (II) or $x > 2$ (III) then $x^2 - 4 > 0$. From (I), (II), and (III), $x > 2$, or graphically:

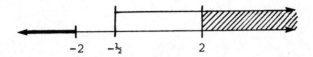

Second case:

$2x + 1 < 0$ and $x^2 - 4 < 0$

$2x + 1 < 0$

$2x < -1$

$x < -\dfrac{1}{2}$ (I)

$x^2 - 4 < 0$

$-2 < x < 2$ (II)

From (I) and (II) $-2 < x < -\dfrac{1}{2}$ or graphically:

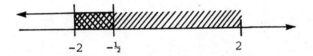

27. **(E)**

The following formula can be used to solve this problem:

$$L = \sqrt{3^2 + 4^2 + 5^2}$$

Using your calculator, you can easily find the answer to be 7.07.

28. **(B)**

A slope of a line is defined as the change rate of y with respect to x. By changing a line equation into the standard form, i.e., the form

$$y = ax + b,$$

you find the slope of the line easily, because it is simply a. Thus, for the given problem:

$$y = -\frac{15}{37}x + \frac{23}{37} = -0.405x + 0.623$$

So, $a = -0.405$.

29. **(B)**

If $(x - 2)$ is a factor, then 2 is a root so $f(2) = 0$. To obtain $f(2)$ we substitute 2 into the expression

$$x^3 - 7x^2 + kx - 12,$$

which results in:

$$2^3 - 7(2) + k(2) - 12 = 8 - 28 + 2k - 12$$

Since $f(2) = 0$, we obtain:

$$8 - 28 + 2k - 12 = 0$$

$$2k - 32 = 0$$

$$2k = 32$$

$$k = 16$$

30. **(D)**

If we sketch the graphs of the functions, we will obtain:

(A)

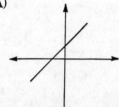

(B)

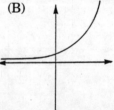

(C)

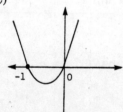

(D)

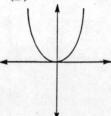

(E)

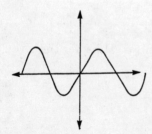

The only graph with symmetry with respect to the y-axis is choice (D).

Note: For a function to be symmetric with respect to the y-axis, $f(x)$ must equal $f(x)$ for all x. The answer to this problem can be determined that way.

31. **(E)**

Let's call the event winning A and not winning $\overline{A}$. We want the probability P given by the expression below:

$$P = P(A) \times P(A) \times P(A) \times P(\overline{A})$$

$$= \frac{3}{4} \times \frac{3}{4} \times \frac{3}{4} \times \frac{1}{4} = \frac{27}{256}$$

32. **(C)**

Consider the trigonometric relationship:

$$\sin^2 x + \cos^2 x = 1$$

By dividing both sides by $\sin^2 x$, we obtain:

$$\frac{\sin^2 x}{\sin^2 x} + \frac{\cos^2 x}{\sin^2 x} = \frac{1}{\sin^2 x}$$
$$1 + \cot^2 x = \csc^2 x$$

33. **(B)**

First, notice that 2.5 is in radian measure. So, set your calculator to the radian mode. Then do the calculation:

$$f(2.5) = \frac{\sin 2.5 \times \cos 5}{\tan 7.5}$$

$$= \frac{0.5985 \times 0.28366}{2.706} = 0.063$$

34. **(A)**

The formula for this problem is

$$D = \frac{|Ax_0 + By_0 + Cz_0 + D|}{\sqrt{A^2 + B^2 + C^2}}$$

where, for this problem, $A = 2$, $B = -3$, $C = 5$, $D = 3$, $x_0 = 2$, $y_0 = 3$, $z_0 = -1$. So, we have

$$D = \frac{|4 - 9 - 5 + 3|}{\sqrt{4 + 9 + 25}} = \frac{7}{\sqrt{38}} = 1.136$$

35.   **(D)**

We employ the slope intercept form for the equation to be written, since we are given the y-intercept. Our task is then to determine the slope.

We are given the equation of a line parallel to the line whose equation we wish to find. We know that the slopes of two parallel lines are equal. Hence, by finding the slope of the given line, we will also be finding the unknown slope. To find the slope of the given equation

$$6x + 3y = 4,$$

we transform the equation $6x + 3y = 4$ into slope intercept form.

$$6x + 3y = 4$$

$$3y = -6x + 4$$

$$y = -\frac{6}{3}x + \frac{4}{3}$$

$$y = -2x + {}^4/_3,$$

Therefore, the slope of the line we are looking for is -2. The y-intercept is -6. Applying the slope intercept form,

$$y = mx + b,$$

to the unknown line, we obtain,

$$y = -2x - 6$$

as the equation of the line.

36.   **(A)**

First, the property of logarithm

$$\log x^{-p} = -p \log x$$

will have to be used. Another property that you need is

$$x^{-p} = \frac{1}{x^p}.$$

With these two properties, you can solve this problem with ease:

$$f(2) = 2^{\log 2^{-3}} = 2^{-3 \log 2}$$

$$= \frac{1}{2^{3 \log 2}} = \frac{1}{2^{0.903}} = \frac{1}{1.86995}$$

$$= 0.535$$

37.   **(E)**

If the vertex is (2, 4) we can substitute this value into the equation of the parabola and obtain:

$$4 = a(2)^2 + b(2) + 3$$

$$4 = \quad 4a + 2b + 3$$

$$\underline{-3 = \qquad\qquad -3}$$

$$1 = \quad 4a + 2b \qquad\qquad\qquad \text{(I)}$$

Also the x-coordinate of the vertex is given by

$$\frac{-b}{2a}$$

so $$2 = \frac{-b}{2a} \qquad\qquad\qquad \text{(II)}$$

From (I) and (II) we obtain

$$4a + 2b = 1 \qquad\qquad\qquad \text{(I)}$$

$$\frac{-b}{2a} = 2 \qquad\qquad\qquad \text{(II)}$$

From (II), $b = -4a$ and substituting into (I):

$$4a + 2 \times (-4a) = 1$$

$$4a - 8a = 1$$

$$-4a = 1$$

$$a = -\frac{1}{4}$$

$$b = 1$$

38.   **(E)**

The given equation has a factor of $(x - 1.5)$, which means it has a root of 1.5. This is also a solution of the equation. Therefore, after replacing x with 1.5, the equation should still hold:

$$1.5^4 + a(1.5)^3 + 2 \times (1.5)^2 + 6 \times (1.5) + 8 = 0$$

Thus,

$$-a = \frac{1.5^4 + 2 \times (1.5)^2 + 6 \times (1.5) + 8}{1.5^3}$$

Use your calculator:

$$a = -\frac{5.0625 + 4.5 + 9 + 8}{3.375} = -7.87$$

39.   **(D)**

To solve this question we need to remember the formulas for the sum of the terms of an arithmetic and a geometric progression.

The sum of the arithmetic progression equals

$$S_{AP} = \frac{n}{2}\left[2u_1 + (n-1)r\right] \tag{I}$$

The sum of the geometric progression equals

$$S_{GP} = \frac{u_1\left(1 = r^n\right)}{1-r} \tag{II}$$

Since $r = 2$ and

$$\frac{S_{AP}}{S_{GP}} = \frac{2}{3},$$

by substitution we obtain:

$$\frac{2}{3} = \frac{\frac{4}{2}\left[2 \times u_1 + (4-1) \times 2\right]}{\frac{u_1\left(1-2^4\right)}{1-2}}$$

$$= \frac{4u_1 + 12}{15u_1}$$

By cross-multiplication we obtain:

$$12u_1 + 36 = 30u_1$$

$$36 = 18u_1$$

$$u_1 = 2$$

Substituting u_1 into (I) and (II), we obtain:

$$S_{AP} = \frac{4}{2}\left[2 \times 2 + (4-1) \times 2\right]$$

$$= \frac{4}{2}(4+6) = 20$$

$$S_{GP} = \frac{2\left(1-2^4\right)}{1-2}$$

$$= \frac{2 \times (-15)}{-1} = 30$$

Therefore $S_{AP} + S_{GP} = 20 + 30 = 50$.

40. **(A)**

$$\frac{\sqrt[3]{-16}}{\sqrt[3]{-2}} = \frac{\sqrt[3]{(-2)8}}{\sqrt[3]{-2}}$$

$$= \frac{\sqrt[3]{-2}\,\sqrt[3]{8}}{\sqrt[3]{-2}}$$

$$= \sqrt[3]{8} = 2$$

Note that since 3 is an odd number, $\sqrt[3]{-2}$ is defined to be a real number and is not imaginary.

41. **(B)**

$$f(x, y) = x^2 + 2y^2 - y \text{ and } g(h) = h + 7.$$

$$g(-2) = -2 + 7 = 5$$

$$f(3, g(-2)) = 3^2 + 2(g(-2))^2 - g(-2)$$

$$= 3^2 + 2(5)^2 - (5)$$

$$= 9 + 50 - 5$$

$$= 54.$$

42. **(B)**

The required set is the set of all x such that

$$\frac{2}{x+1} - 3 = \frac{4x+6}{x+1}$$

Multiplying each member by $(x + 1)$ to eliminate the fraction we obtain

$$(x+1)\left(\frac{2}{x+1} - 3\right) = \left(\frac{4x+6}{x+1}\right)(x+1)$$

Distributing,

$$(x+1)\left(\frac{2}{x+1}\right) - (x+1) = 4x+6$$

$$2 - (3x+3) = 4x+6$$

$$-1 - 3x = 4x+6$$

$$-1 - 3x - 4x = 6$$

$$-7x = 7$$

$$x = -1$$

If we now substitute (− 1) for x in our original equation,

$$\frac{2}{x+1} - 3 = \frac{4x+6}{x+1}$$

$$\frac{2}{-1+1} - 3 = \frac{4(-1)+6}{-1+1}$$

$$\frac{2}{0} - 3 = \frac{-4+6}{0}$$

Since division by zero is impossible the above equation is not defined for $x = -1$. Hence we conclude that the equation has no roots and for $x = -1$.

$$\left\{ x \left| \frac{2}{x+1} - 3 = \frac{4x+6}{x+1} \right. \right\} = \phi,$$

where ϕ is the empty set.

43. **(D)**

   A sphere is formed with a circle is rotated 360° about its diameter.

44. **(C)**

$$(f \circ g)(x) = f(x) \circ g(x) = f(g(x)).$$

This is the definition of the composition of functions. For the functions given in this problem we have

$$2(g(x)) + 4 = 2(x^2 - 2) + 4$$

$$= 2x^2 - 4 + 4 = 2x^2.$$

45. **(A)**

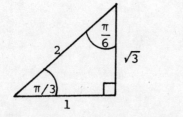 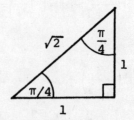

   Express $\cos^{\pi}/_{12}$ in terms of angles whose values of the trigonometric functions are known:

$$\cos\frac{1}{12}\pi = \cos\left(\frac{4}{12}\pi - \frac{3}{12}\pi\right)$$

$$= \cos\left(\frac{1}{3}\pi - \frac{1}{4}\pi\right)$$

Now apply the difference formula for the cosine of two angles, α and β.

$$\cos(\alpha - \beta) = \cos\alpha\,\cos\beta + \sin\alpha\,\sin\beta.$$

In this example,

$$\alpha = \frac{1}{3}\pi \text{ and } \beta = \frac{1}{4}\pi.$$

$$\cos\left(\frac{1}{3}\pi - \frac{1}{4}\pi\right) = \cos\frac{1}{3}\pi\,\cos\frac{1}{4}\pi + \sin\frac{1}{3}\pi\,\sin\frac{1}{4}\pi$$

See the accompanying diagrams to find the values of these angles. We find:

$$\cos\frac{\pi}{3} = \frac{1}{2}$$

$$\cos\frac{\pi}{4} = \frac{1}{\sqrt{2}}$$

$$\sin\frac{\pi}{3} = \frac{\sqrt{3}}{2}$$

$$\sin\frac{\pi}{4} = \frac{1}{\sqrt{2}}$$

Thus:

$$\cos\left(\frac{1}{3}\pi - \frac{1}{4}\pi\right) = \frac{1}{2} \times \frac{1}{\sqrt{2}}\left(\frac{\sqrt{2}}{\sqrt{2}}\right) + \frac{\sqrt{3}}{2} \times \frac{1}{\sqrt{2}}\left(\frac{\sqrt{2}}{\sqrt{2}}\right)$$

$$= \frac{\sqrt{2}}{4} + \frac{\sqrt{2}\sqrt{3}}{4}$$

$$= \frac{1}{4}\sqrt{2}(1 + \sqrt{3})$$

$$= \frac{\sqrt{2}}{4}(1 + \sqrt{3})$$

46.   **(C)**
Note that

$$32 = 2 \times 2 \times 2 \times 2 \times 2 = 2^5.$$

Thus:

$$\log_{10}32 = \log_{10}2^5.$$

Recall the logarithmic property,

$$\log_b x^y = y\log_b x.$$

Hence:

$$\log_{10} 32 = \log_{10} 2^5 = 5\log_{10} 2$$

$$= 5(0.3010)$$

$$= 1.5050$$

47. **(C)**

$$2\cos x - \sqrt{2} = 0$$

$$\cos x = \frac{\sqrt{2}}{2} = \frac{1}{\sqrt{2}}$$

$$x = 45°.$$

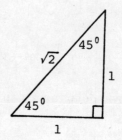

48. **(A)**

Using the cosine difference formula we have

$$\cos\left(\frac{\pi}{2} - \theta\right) = \cos\frac{\pi}{2}\cos\theta + \sin\frac{\pi}{2}\sin\theta$$

$$= 0 \times \cos\theta + 1 \times \sin\theta$$

$$= \sin\theta.$$

49. **(B)**

$$f(x) = x - 2x^2 + 2.$$

Therefore:

$$f(a - 2) = (a - 2) - 2(a - 2)^2 + 2$$

$$= (a - 2) - 2(a^2 - 4a + 4) + 2$$

$$= a - 2 - 2a^2 + 8a - 8 + 2$$

$$= 9a - 2a^2 - 8.$$

50. **(A)**

In order to find the inverse of the function, we interchange x and y and solve for y. We have

$$y^2 = x, y = \pm\sqrt{x},$$

but one of our conditions is that $x \geq 0$, hence

$$f^{-1} = \{(x, y) : y = +\sqrt{x}\}.$$

THE SAT SUBJECT TEST IN

Math
Level 2

PRACTICE TEST 2

This test is also on CD-ROM in our special interactive SAT Math Level 2 TEST*ware*®. It is highly recommended that you first take this exam on computer. You will then have the additional study features and benefits of enforced timed conditions and instant, accurate scoring. See page 2 for guidance on how to get the most out of our SAT Math Level 2 software.

SAT Mathematics Level 2

Practice Test 2

Time: 1 Hour
50 Questions

DIRECTIONS: Choose the best answer for each question and mark the letter of your selection on the corresponding answer sheet in the back of the book.

NOTES:

(1) Some questions require the use of a calculator. You must decide when the use of your calculator will be helpful.

(2) You may need to decide which mode your calculator should be in—radian or degree.

(3) All figures are drawn to scale and lie in a plane unless otherwise stated.

(4) The domain of any function f is the set of all real numbers x for which $f(x)$ is a real number, unless other information is provided.

REFERENCE INFORMATION: The following information may be helpful in answering some of the questions.

Volume of a right circular cone with radius r and height h

$$V = \frac{1}{3}\pi r^2 h$$

Lateral area of a right circular cone with circumference of the c and slant height l

$$S = \frac{1}{2}cl$$

Volume of a sphere with radius r

$$V = \frac{4}{3}\pi r^3$$

Surface Area of a sphere with r

$$S = 4\pi r^2$$

Volume of a pyramid with base area B and height h

$$V = \frac{1}{3}Bh$$

1. If $b^2 - 4c = 0$, where b and c are real numbers, then the roots of the equation

   $$x^2 + bx + c = 0 \text{ are}$$

   (A)   real and equal.

   (B)   real and unequal.

   (C)   complex and equal.

   (D)   complex and unequal.

   (E)   No solution.

2. Find $2^{\frac{3}{4}} + 2^{\frac{4}{3}}$.

   (A)   4.2

   (B)   1.414

   (C)   8.5

   (D)   6

   (E)   2.3

3. If f is defined by

   $$f(x) = \frac{5x - 8}{2}$$

   for each real number x, find the solution set for $f(x) > 2x$.

   (A) $\{x \mid x > 6\}$

   (B) $\{x \mid x > 8\}$

   (C) $\{x \mid x < 8\}$

   (D) $\{x \mid 6 < x < 8\}$

   (E)   None of the above.

4. If $f(x) = \sqrt{x^3 + 5}$, then what is $f^{-1}(0.5)$?

   (A) -1.68

   (B)   1.68

   (C)   2.5

   (D) -2.5

   (E)   2.26

5. If two fair dice are thrown, what is the probability that the sum of the number of dots on the top faces will be 7?

   (A) $\dfrac{1}{2}$

   (B) $\dfrac{1}{6}$

(C) $\dfrac{1}{9}$ (D) $\dfrac{1}{12}$

(E) $\dfrac{1}{16}$

6.   If

$$\frac{2\sin\dfrac{5\pi}{6}+\cos x}{\tan\dfrac{\pi}{2}+\cos\dfrac{2\pi}{3}}=0$$

and $0 \le x \le \pi$, then $x =$

(A) $\dfrac{\pi}{2}$ (D) π

(B) $\dfrac{2\pi}{3}$ (E) $\dfrac{5\pi}{6}$

(C) $-\dfrac{\pi}{2}$

7.   If $f(x) = 3x^2$ and $f(g(x)) = 25x$, then $g(x) =$

(A) $\pm\,2.89\sqrt{x}$ (D) $-2.89\sqrt{x}$

(B) $2.45x$ (E) $2.45\sqrt{x}$

(C) $2.89\sqrt{x}$

8.   The graph below represents a parabola of the form

   $ax^2 + bx + c = y.$

Which of the statements below is true?

(A) $c = 0$

(B) $b^2 > 4a$

(C) $b^2 < 4a$

(D) $c < 0$

(E) $b = 2\sqrt{ac}$

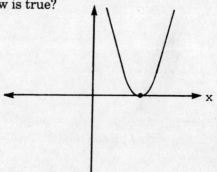

9. A cylinder closed at both ends has a volume of 18π cm^3. If the ratio between its height and its radius is 2:3, what is the surface area of the cylinder?

    (A) 21π cm^2

    (B) 30π cm^2

    (C) 27π cm^2

    (D) 19π cm^2

    (E) 5π cm^2

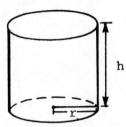

10. The equation

    $$x^2 + 2(k + 2)x + 9k = 0$$

    has equal roots. k is therefore equal to

    (A)   4 only.   (D) -1 and 4.

    (B)   1 and 4.   (E)   2 and -4.

    (C)   0 and 4.

11. The lengths of the sides of a triangle are 5, 4, and 7. What is the largest angle of this triangle?

    (A)   36.3°   (D)   101.5°

    (B)   165.2°   (E)   97.1°

    (C)   57°

12. Find the intersection of the parabola $y = 2x^2$ and its inverse.

    (A) $(0, 0)$

    (B) $(0, 0), (-1/2, 1/2), (-1/2, -1/2)$

    (C) $(0, 0), (1/2, 1/2), (-1/2, 1/2)$

    (D) $(0, 0), (-1/2, 1/2)$

    (E) $(0, 0), (1/2, 1/2)$

13. The function f is defined by

$$f(x) = \frac{1}{1+x}.$$

For what values of x is $f(f(x))$ undefined?

(A) {0}

(D) $\{-1, -\frac{1}{2}\}$

(B) {−1, 0}

(E) {−1, −2}

(C) $\{-\frac{1}{2}, 0\}$

14. Which of the following relations has an inverse which is not a function?

I. $y = \sqrt{x}$

II. $|y| = x$

III. $y = \sin x$

(A) I only.

(D) I and II.

(B) II only.

(E) I and III.

(C) III only.

15. A right cone is inscribed in a hemisphere of radius r such that the base of the cone coincides with the base of the hemisphere. What is the ratio of the volume of the hemisphere to the volume of the cone?

(A) 3:2

(D) 3:1

(B) 2:1

(E) 1:2

(C) 4:3

16. If $0 < x < \frac{\pi}{2}$ and $\tan 7x = 3$, what is the value of $\tan x$?

(A) 0.5

(D) 0.28

(B) 0.4

(E) 0.18

(C) 0.38

17. How many four-digit numbers are there such that the first digit is odd, the second is even, and there is no repetition of digits?

(A) 1,200

(D) 1,400

(B) 1,625

(E) 1,600

(C) 200

18. If $f(x, y) = \dfrac{\log x}{\log y}$ then $f(4, 2) =$

(A) 0

(D) 2

(B) $\dfrac{1}{2}$

(E) log2

(C) 1

19. What is the value of b in the relation $\log_b \dfrac{1}{25} = -2$?

(A) 2

(D) 25^2

(B) $\pm\sqrt{2}$

(E) ± 5

(C) 2^{25}

20. What are the truth values for r, s, t, and u, respectively?

p	q	$p \to q$	$q \to p$	$(p \to q) \wedge (q \to P)$
T	T	T	(t)	T
T	(r)	F	T	(u)
F	T	(s)	F	F
F	F	T	T	T

(A) F, T, T, F

(D) T, T, F, F

(B) F, F, T, F

(E) F, T, T, T

(C) T, T, T, F

21. Determine the values of $\sin x$ if $\sin x = \cos 2x$.

(A) $\pm\dfrac{\sqrt{3}}{2}, 0$

(D) $-1, \dfrac{1}{2}$

(B) $\pm\dfrac{\sqrt{2}}{2}$

(E) $0, \dfrac{1}{2}$

(C) $-\dfrac{1}{2}, -1$

22. If $f(x) = 1 + 3x - 2kx^2$ and $f(-\dfrac{1}{2}) = 0$ then $k =$

(A) 0

(D) -1

(B) $\dfrac{1}{2}$

(E) 5

(C) 1

23. Express the equation $x^2 + xy + y^2 = 1$ in polar coordinates r and θ.

(A) $r^2 + \cos\theta \sin\theta = 1$

(B) $r^2 + r\cos\theta \sin\theta = 1$

(C) $r^2 \sin^2\theta + r\cos\theta \sin\theta = 1$

(D) $r^2(1 + \cos\theta \sin\theta) = 1$

(E) $r^2 + \sin^2\theta = 1$

24. Let $f : R \rightarrow R$ and $g : R \rightarrow R$ be two functions given by

$f(x) = 2x + 5$ and $g(x) = 4x^2$

respectively for all x in R, where R is the set of real numbers. Find expressions for the composition $(f \circ g)(x)$.

(A) $8x^3 + 20x^2$

(D) $8x^3 + 5$

(B) $4(2x + 5)^2$

(E) None of the above.

(C) $8x^2 + 5$

25. A point has rectangular coordinates (5,3). The point has polar coordinates

    (A)   (5.8, 31°)   (D)   28°

    (B)   (4.2, 28°)   (E)   4.2

    (C)   5.8

26. Find x if $\log 100 - \log \dfrac{1}{5} = 2 + \log x$, where $\log x$ means $\log_{10} x$.

    (A)   3   (D)   6

    (B) -1 (E) -3

    (C)   5

27. Find the solution set of

    $$\frac{|x+2|}{|x-1|} = 3.$$

    (A) $\{\dfrac{5}{2}, \dfrac{1}{4}\}$ (D) $\{\dfrac{1}{2}, 1\}$

    (B) $\{-\dfrac{2}{5}, -\dfrac{1}{2}\}$ (E) $\{\dfrac{1}{2}, 2\}$

    (C) $\{-2, 2\}$

28. Given the table below, how much of X and Y should be combined to obtain 60 liters of a solution 3.2% in ethanol?

Solution (liters)	Percent of ethanol
X	2
Y	6

    (A) $X = 18, Y = 42$ (D) $X = 40, Y = 20$

    (B) $X = 20, Y = 40$ (E) $X = 42, Y = 18$

    (C) $X = 25, Y = 35$

29. If

$$\sum_{k=0}^{5}(2 + 3x + k) = P + \sum_{k=0}^{5} k$$

and equals 45 when $x = 1$, what is P when $x = 2$?

(A)   40   (D)   48

(B)   8   (E)   5

(C)   42

30. If $3 \arctan \dfrac{1}{\sqrt{3}} = \arcsin x$ then $x =$

(A)   1

(B)   0

(C) $\dfrac{1}{2}$

(D) $\dfrac{\sqrt{3}}{2}$

(E) $\dfrac{1}{\sqrt{2}}$

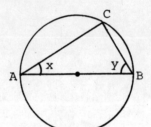

31. $\log_3\left(\dfrac{1}{27}\right) =$

(A)   −3   (D)   3

(B) $-\dfrac{1}{3}$ (E)   9

(C) $\dfrac{1}{3}$

32. If the functions f and g are defined for all x by

$f(x) = 2x^2 - 3x + 2$ and $g(x) = 2$ then $g(f(1)) =$

(A)   0   (D)   3

(B)   1   (E)   4

(C)   2

33. If $\sin x = \dfrac{1}{2}$ and $0 \le x \le \dfrac{\pi}{2}$ then

$$\sin\left(x + \dfrac{\pi}{4}\right) =$$

(A) $\dfrac{\sqrt{2}(\sqrt{3} - 1)}{2}$

(D) $\sqrt{2}(\sqrt{3} + 1)$

(B) $\dfrac{\sqrt{2}}{2}(1 + \sqrt{3})$

(E) $\dfrac{\sqrt{2}}{4}(1 + \sqrt{3})$

(C) $\sqrt{2}\,(\sqrt{3} - 1)$

34. $(\sin\theta \times \cot\theta)^2 + (\cos\theta \times \tan\theta)^2 =$

(A) 1

(D) $2\cot^2\theta$

(B) $2\cos^2\theta$

(E) $2\tan^2\theta$

(C) $2\sin^2\theta$

35. If y varies inversely as the square of x, what is the effect on x when y is divided by 4?

(A) multiplied by 2

(D) divided by 4

(B) multiplied by 4

(E) multiplied by $\dfrac{3}{2}$

(C) divided by 2

36. $\log_3\left(\dfrac{1}{81}\right) =$

(A) -4

(D) $-\dfrac{1}{4}$

(B) $\dfrac{1}{4}$

(E) 3

(C) 4

37. A circle

$$x^2 + 4x + y^2 + 4y = 0$$

has an equilateral triangle inscribed inside it. Find the length of the side of this triangle.

(A)   3.0

(B)   4.9

(C)   2.5

(D)   1.8

(E)   6.2

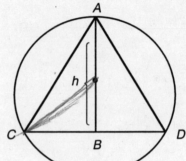

38. Simplify $\dfrac{\sqrt{-20}}{\sqrt{-5}}$.

(A) $i\sqrt{4}$

(B) $\sqrt{2i}$

(C) $\dfrac{\sqrt{4}}{\sqrt{5}}i$

(D)   2

(E) $\dfrac{i}{\sqrt{2}}$

39. The equations of translation from the $x_1 - y_1$ plane to the $x-y$ plane are $x = x_1 - 4$, $y = y_1 + 2$. Find the equation of the line l:

$$2x + 3y - 18 = 0$$

when referred to the $x_1 - y_1$ plane.

(A) $x_1 + y_1 - 2$

(B) $x_1 - 4 + y_1 - 2$

(C) $2x_1 + 3y_1 - 20 = 0$

(D) $x_1 - y_1 - 6 = 0$

(E) $x_1 + y_1 \rightarrow \infty$

40. Two cars travel at 40 and 60 miles per hour, respectively. If the second car starts out 5 miles behind the first car, how many hours will it take the second car to overtake the first car?

(A) $\dfrac{1}{2}$ (D) $\dfrac{1}{5}$

(B) $\dfrac{1}{4}$ (E) Cannot be determined.

(C) 1

41. Which of the following is a root for the equation

$$-2x^2 + 4x + 5 = 0?$$

(A) 2.15   (D) 5.2

(B) 3.48   (E) 3.22

(C) 2.87

42. Solve the following inequality:

$$4(2x - 6) - 10x \le -28$$

(A) $x < 2$ (D) $x \ge 26$

(B) $x > 2$ (E) $x \le 26$

(C) $x \ge 2$

43. The roots of an equation are 2, –3, and $\dfrac{7}{5}$. What is the equation?

(A) $5x^2 - 37x + 21 = 0$

(B) $5x^3 - 2x^2 - 37x + 42 = 0$

(C) $5x^3 + 2x^2 + 37x - 42 = 0$

(D) $2x^3 - 2x^2 + 37x - 21 = 0$

(E) Cannot be determined.

44. The distance between the point (2, 4) and the line

$$y + 2x - 3 = 0 \text{ is}$$

(A) 1.9   (D) 3.8

(B)   3.24   (E)   5.2

(C)   2.24

45.   If $\cos 35° = \tan x°$, then $x =$

(A)   0.686   (D)   19.5

(B)   39.32   (E)   79.3

(C)   27

46.   What is the value of $(0.0081)^{-\frac{3}{4}}$?

(A) $\dfrac{81}{27}$ (D) $\dfrac{8.10 \times 10^{-2}}{4}$

(B) $\dfrac{1,000}{27}$ (E) $\dfrac{8.10 \times 10^{-2}}{0.27}$

(C) $\dfrac{27}{81}$

47.   A sphere having surface area 10 must have a volume of

(A)   13.34   (D)   2.97

(B)   6.83   (E)   3.45

(C)   1.85

48.   Two fair dice are tossed 5 times. Find the probability that 7 will show on the first three tosses and will not show on the other two.

(A) $\dfrac{1}{6}$ (D) $\left(\dfrac{1}{6}\right)^3\left(\dfrac{5}{6}\right)^2$

(B) $\dfrac{2}{3}$ (E) $3\left(1-\dfrac{1}{6}\right)$

(C) $\left(\dfrac{5}{6}\right)\left(\dfrac{1}{6}\right) \times 3$

49. Which of the following vectors is equal to **MN** if $M = (2, 1)$ and $N = (3, -4)$?

    (A) **AB**, where $A = (1, -1)$ and $B = (2, 3)$

    (B) **CD**, where $C = (-4, 5)$ and $D = (-3, 10)$

    (C) **EF**, where $E = (3, -2)$ and $F = (4, -7)$

    (D) **GH**, where $G = (3, 2)$ and $H = (7, -4)$

    (E) **KL**, where $K = (-2, 3)$ and $L = (-7, 4)$

50. Given the quadrilateral *ABCD* with vertices at $A(-3, 0)$, $B(9, 0)$, $C(9, 9)$, $D(0, 12)$, find the area of quadrilateral *ABCD*.

    (A) 94.5

    (B) 112.5

    (C) 99

    (D) 105

    (E) Cannot be determined.

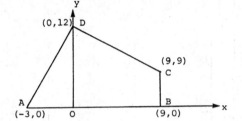

SAT Mathematics Level 2

Practice Test 2
ANSWER KEY

1.	(A)	14.	(C)	27.	(A)	40.	(B)
2.	(A)	15.	(B)	28.	(E)	41.	(C)
3.	(B)	16.	(E)	29.	(D)	42.	(C)
4.	(A)	17.	(D)	30.	(A)	43.	(B)
5.	(B)	18.	(D)	31.	(A)	44.	(C)
6.	(D)	19.	(E)	32.	(C)	45.	(B)
7.	(A)	20.	(A)	33.	(E)	46.	(B)
8.	(E)	21.	(D)	34.	(A)	47.	(D)
9.	(B)	22.	(D)	35.	(A)	48.	(D)
10.	(B)	23.	(D)	36.	(A)	49.	(C)
11.	(D)	24.	(C)	37.	(B)	50.	(B)
12.	(E)	25.	(A)	38.	(D)		
13.	(E)	26.	(C)	39.	(C)		

DETAILED EXPLANATIONS
OF ANSWERS

1.　　**(A)**
If $x^2 + bx + c = 0$, then by the quadratic formula

$$x = \frac{-b \pm \sqrt{b^2 - 4c}}{2}.$$

We are given $b^2 - 4c = 0$, so the two roots

$$x_1 = \frac{-b+0}{2} \text{ and } x_2 = \frac{-b-0}{2}.$$

So x_1 and x_2 are real numbers and $x_1 = x_2$.

2.　　**(A)**
Simplifying power terms into more familiar form is the first step for this problem. This is done like this:

$$x^{\frac{q}{p}} = \sqrt[p]{x^q}$$

So, the problem can be put into a simpler form of the following:

$$\sqrt[3]{2^4} + \sqrt[4]{2^3} = \sqrt[3]{16} + \sqrt[4]{8}$$

which equals 4.2.

3.　　**(B)**
To find the solution set of $f(x) > 2x$, we proceed as follows:

$$\frac{5x-8}{2} > 2x$$

$$5x - 8 > 4x$$

which implies $x > 8$.

4.　　**(A)**
First, find the inverse function of $f(x)$. The way to do this is to assume $f(x) = y$, where y is another variable. Thus for each value of x, there is a value of y,

which equals $f(x)$. Solve for x in terms of y, which will give you exactly $f^{-1}(y)$. That is

$$f^{-1}(y) = \sqrt[3]{y^2 - 5}$$

Put the value of 0.5 into this function. We have

$$f^{-1}(0.5) = \sqrt[3]{0.5^2 - 5} = -1.68$$

5.　　**(B)**
There are 36 possible outcomes. Six of these produce the number 7:

(1, 6), (2, 5), (3, 4), (4, 3), (5, 2), and (6, 1).

Hence the probability is

$$\frac{6}{36} = \frac{1}{6}.$$

6.　　**(D)**
We need

$$2\sin\frac{5\pi}{6} + \cos x = 0.$$

$$\sin\frac{5\pi}{6} = \frac{1}{2},$$

hence $\cos x = -1 \Rightarrow x = \pi$.

7.　　**(A)**
$f(g(x))$ is a function $g(x)$, so we can plug $g(x)$ into it:

$$f(g(x)) = 3(g(x))^2$$

But $f(g(x)) = 25x$, therefore,

$$3(g(x))^2 = 25x$$

or $g(x) = \pm\sqrt{\dfrac{25}{3}}\ \sqrt{x}$.

Calculator:　　$\sqrt{25 \div 3} = \sqrt{8.3333} = 2.88675$.

8.　　**(E)**
The x-coordinate of the vertex is given by

$\dfrac{-b}{2a}$.

If we substitute this value into

$$y = ax^2 + bx + c,$$

we obtain:

$$y = a\left(\frac{-b}{2a}\right)^2 + b\left(\frac{-b}{2a}\right) + c$$

$$y = a\left(\frac{b^2}{4a^2}\right) - \frac{b^2}{2a} + c$$

$$y = \frac{b^2}{4a} - \frac{b^2}{2a} + c$$

$$y = \frac{-b^2}{4a} + c$$

Since $y = 0$ at the vertex (from the graph)

$$\frac{-b^2}{4a} + c = 0 \ \therefore c = \frac{b^2}{4a} \text{ and } b = 2\sqrt{ac}$$

9.   **(B)**

The volume of a cylinder is given by:

$$V = \pi r^2 h$$

Since $\dfrac{h}{r} = \dfrac{2}{3} \Rightarrow h = \dfrac{2}{3}r$

$$\pi r^2\left(\frac{2}{3}r\right) = 18\pi$$

$$\frac{2}{3}r^3 = 18$$

$$r^3 = 27 \Rightarrow r = 3 \text{ and } h = \frac{2}{3} \times 3 = 2$$

The surface area is given by

$$A_s = 2\pi r^2 + 2\pi rh$$

$$= 2\pi \times 9 + 2\pi \times 3 \times 2$$

$$= 18\pi + 12\pi = 30\pi$$

10.   **(B)**

The given equation is a quadratic equation of the form

$$ax^2 + bx + c = 0.$$

In the given quadratic equation, $a = 1$, $b = 2(k + 2)$, and $c = 9k$. A quadratic equation has equal roots if the discriminant, $b^2 - 4ac$, is zero.

$$b^2 - 4ac = [2(k + 2)]^2 - 4(1)\,(9k) = 0$$

$$4(k + 2)^2 - 36k = 0$$

$$4(k + 2)\,(k + 2) - 36k = 0$$

$$4(k^2 + 4k + 4) - 36k = 0$$

Distributing, $4k^2 + 16k + 16 - 36k = 0$

$$4k^2 - 20k + 16 = 0$$

Divide both sides of this equation by 4:

$$\frac{4k^2 - 20k + 16}{4} = \frac{0}{4}$$

or $k^2 - 5k + 4 = 0$

Factoring the left side of this equation into a product of two polynomials:

$$(k - 4)\,(k - 1) = 0$$

When the product $ab = 0$, where a and b are any two numbers, either $a = 0$ or $b = 0$. Hence, in the case of this problem, either

$$k - 4 = 0 \text{ or } k - 1 = 0$$

Therefore, $k = 4$ or $k = 1$

11.   **(D)**

The angle opposite the side with length 7 has the largest angle. Following the law of cosines,

$$7^2 = 5^2 + 4^2 - 2 \times 5 \times 4 \times \cos\theta$$

where θ is that angle. So, then

$$\theta = \cos^{-1}\frac{49 - 25 - 16}{-2 \times 5 \times 4} = 101.5$$

12.   **(E)**

The equation of the parabola is $y = 2x^2$ and its inverse is given by $x = 2y^2$. The ordered pairs that satisfy both at the same time correspond to their intersection:

$$\begin{cases} y = 2x^2 & \text{(I)} \\ x = 2y^2 & \text{(II)} \end{cases}$$

If we substitute (II) into (I) we will obtain:

$$y = 2(2y^2)^2$$

$$y = 8y^4$$

$$y - 8y^4 = 0$$

$$y(1 - 8y^3) = 0 \Rightarrow y = 0 \text{ or } y = \frac{1}{2}$$

If $y = 0$, then $x = 0$

If $y = \frac{1}{2}$, from (I) $x = \pm\frac{1}{2}$.

Note that the ordered pair $(-1/2, 1/2)$ has to be eliminated, since it does not satisfy (II). The solution consists of the pairs $(0, 0)$ and $(1/2, 1/2)$.

13.   **(E)**

$$f(f(x)) = \frac{1}{1 + \dfrac{1}{1+x}},$$

which is undefined for

$$1 + x = 0 \text{ and } 1 + \frac{1}{1+x} = 0.$$

That is, $f(f(x))$ is undefined for $x = -1$ and $x = -2$.

14.   **(C)**

The domain of $y = \sqrt{x}$ is

$$\{x \in R \mid x \geq 0\};$$

the range is also the same. The inverse of $x = \sqrt{y}$ or $y = x^2$, $x \geq 0$. This, of course, is a function. The relation $\mid y \mid = x$ is not a function; each x value, for example $x = 1$, has two corresponding y values, $y = \pm 1$. However, the inverse

$$y = \mid x \mid, x \in R,$$

is a function. The function $y = \sin x$ is periodic; the value assigned at x is the same as that assigned at

$$x + 2n\pi, \; n = 0, \pm 1, \pm 2, \ldots$$

So the inverse will assign the same x values to y and $y + 2n\pi$. The y value must be unique to define a function, so the inverse is not a function. The curves below show each of the relations and their inverses.

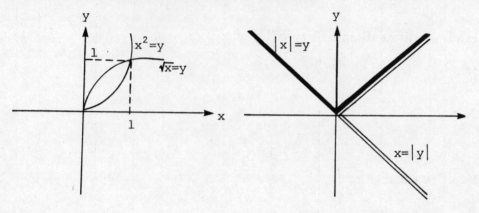

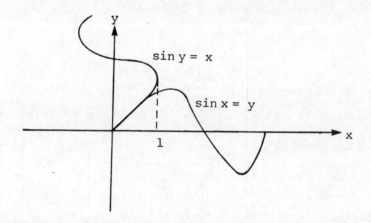

15.   **(B)**

The cone is right, which means that the line passing through the center of the base and perpendicular to it also passes through the vertex of the cone. Therefore, the height of the cone must be equal to the radius of the hemisphere. The volume of the cone equals

$$\frac{1}{3} \pi r^3 h = \frac{1}{3} \pi r^3.$$

The volume of a sphere equals

$$\frac{4}{3} \pi r^3,$$

so the volume of the hemisphere is

$$\frac{1}{2}\left(\frac{4}{3}\pi r^3\right) = \frac{2}{3}\pi r^3.$$

Thus, we have

$$\frac{\text{volume of hemisphere}}{\text{volume of cone}} = \frac{\frac{2}{3}\pi r^3}{\frac{1}{3}\pi r^3} = \frac{2}{1}.$$

16.   **(E)**
Assume $y = 7x$. Then $\tan 7x = \tan y = 3$. Thus

$$y = \tan^{-1}(3) = 71.56$$

So, $x = \dfrac{y}{7} = \dfrac{71.56}{7} = 10.22$

Then $\tan x = \tan 10.22 = 0.18$

17.   **(D)**
There are five possibilities for the first (1, 3, 5, 7, 9), five for the second (0, 2, 4, 6, 8), 8 possibilities for the third $(10 - 2)$ and 7 for the fourth $(8 - 1) = 7$. We subtracted 2 possibilities on the third and 3 on the fourth because there is no repetition, as shown below:

$$5 \times 5 \times (10 - 2)(8 - 1) = 5 \times 5 \times 8 \times 7 = 1,400$$

18.   **(D)**

$$f(4, 2) = \frac{\log 4}{\log 2} = \frac{\log 2^2}{\log 2} = \frac{2\log 2}{\log 2} = 2$$

19.   **(E)**
Since the statement,

$$\log_b \frac{1}{25} = -2$$

is equivalent to

$$b^{-2} = \frac{1}{25}$$

and

$$b^{-2} = \frac{1}{b^2}$$

Therefore

$$\frac{1}{b^2} = \frac{1}{25}$$

Cross multiply to obtain the equivalent equation, $b^2 = 25$. Take the square root of both sides. To find $b = \pm 5$.

20.   **(A)**

The sign "→" means conditional and "∧" means conjunction; $p \to q$ should be read "if p then q" and $(p \to q) \wedge (q \to p)$ should be read "if p then q and if q then p."

The truth table for conditionals is represented below:

p	q	$p \to q$
T	T	T
T	F	F*
F	T	T
F	F	T

*Note that a conditional is false if and only if a true hypothesis leads to a false conclusion.

r: Since a conditional is false only if a true p leads to a false q, we conclude that r has to be false.

s: If p is false there is no way to say the conditional is false, so we have to accept it as true.

t: A conjunction is true only if both statements are true, in this case, $(q \to p)$ has to be true.

u: Since $(p \to q)$ is false and both $(p \to q)$ and $(q \to p)$ need to be true for the conjunction to be true, the conjunction is false.

21.   **(D)**

$$\sin x = \cos 2x = \cos^2 x - \sin^2 x$$

$$\sin x = 1 - \sin^2 x - \sin^2 x$$

$$\sin x = 1 - 2\sin^2 x$$

$$0 = -2\sin^2 x - \sin x + 1$$

$$\sin x = \frac{-(-1) \pm \sqrt{(-1)^2 - 4(-2)(1)}}{2(-2)}$$

$$= \frac{1 \pm 3}{-4} = \begin{cases} -1 \\ +\frac{1}{2} \end{cases}$$

22.   **(D)**

$$f\left(-\frac{1}{2}\right) = 1 + 3\left(-\frac{1}{2}\right) - 2k\left(-\frac{1}{2}\right)^2 = 0,$$

$$1 - \frac{3}{2} - 2k\left(\frac{1}{4}\right) = 0$$

$$\Rightarrow \quad \frac{k}{2} = -\frac{1}{2} \text{ or } k = -1$$

23.   **(D)**

Substitute $x = r\cos\theta$, $y = r\sin\theta$ into

$$x^2 + xy + y^2 = 1.$$

$$r^2\cos^2\theta + (r\cos\theta)(r\sin\theta) + r^2\sin^2\theta = 1,$$

$$r^2(\cos^2\theta + \sin^2\theta) + r^2\cos\theta\sin\theta = 1,$$

$$r^2(1) + r^2\cos\theta\sin\theta = 1,$$

$$r^2(1 + \cos\theta\sin\theta) = 1.$$

Note that this is the equation of an ellipse whose main axis makes an angle of 45° with the coordinate axis. A quadratic equation, consisting of the terms x^2, y^2 and xy in addition to first order terms, represents either an ellipse or a hyperbola whose main axes are not necessarily parallel to the coordinate axes. In this example, we have:

$$1 + \sin\theta\cos\theta = 1 + \frac{\sin 2\theta}{2}$$

and $-1 \le \sin 2\theta \le 1 \Rightarrow \dfrac{1}{2} \le (1 + \dfrac{1}{2}\sin 2\theta) \le \dfrac{3}{2}$

$$\Rightarrow \quad \frac{2}{3} \le r^2 = \frac{1}{\left(1 + \dfrac{1}{2}\sin 2\theta\right)} \le 2$$

which implies that

$$\frac{\sqrt{2}}{\sqrt{3}} \le r \le \sqrt{2}.$$

Therefore the curve cannot be a hyperbola because r would have to be boundless. Thus it is an ellipse. (Note that a circle is considered as a special case of an ellipse.)

24. **(C)**

Consider functions $f : A \to B$ and $g : B \to C$, that is, where the codomain of f is the domain of g. Then the function $g \circ f$ is defined as $g \circ f : A \to C$ where

$$(g \circ f)(x) = g(f(x))$$

for all x in A, and it is called the composition of f and g.

In another notation,

$$g \circ f = \{(x, z) \in A \times C \mid \text{for all } y \in B \text{ such that}$$

$$(x, y) \in f \text{ and } (y, z) \in g\}.$$

So $\qquad (f \circ g)(x) = f(g(x))$

$$= f(4x + 2)$$

$$= 2[4x^2] + 5$$

$$= 8x^2 + 5.$$

25. **(A)**

The polar coordinates are characterized by (r, θ), where r is the distance between the point and the origin and θ is the angle rotated away from the point direction of the x-axis. Therefore, we can obtain the value of r as

$$r = \sqrt{5^2 + 3^2} = 5.8$$

The angle can be calculated by

$$\theta = \tan^{-1}\frac{3}{5} = 31°$$

26. **(C)**

$$\log 100 - \log \frac{1}{5} = 2 + \log x$$

$$\log \frac{100}{\frac{1}{5}} = 2 + \log x$$

$$\log 500 = 2 + \log x$$

$$\log(5 \times 10^2) = 2 + \log x$$

$$(\log 5) + 2 = 2 + \log x$$

$$x = 5$$

27. **(A)**

First we note that for any two real numbers we have:

$$\frac{|a|}{|b|} = \left|\frac{a}{b}\right|$$

Hence if in the above equality, we put $a = x + 2$, $b = x - 1$, we will have:

$$\frac{|x+2|}{|x-1|} = \left|\frac{x+2}{x-2}\right| = 3$$

On the other hand, according to the definition of the absolute value of a number, we know that if $|y| = b > 0$, then y can be either b or $-b$.

Hence, setting

$$y = \frac{x+2}{x-1} \text{ and } b = 3$$

in the above equality, we have:

$$\frac{x+2}{x-1} = 3 \text{ or } \frac{x+2}{x-1} = -3.$$

From the first equation we obtain:

$$x + 2 = 3(x - 1)$$

$$5 = 2x$$

$$x = \frac{5}{2}$$

From the second we have:

$$x + 2 = -3(x - 1)$$

$$-1 = -4x$$

$$x = \frac{1}{4}$$

The solution set is thus $\left\{\frac{5}{2}, \frac{1}{4}\right\}$.

28. **(E)**

We want to obtain 60 liters of solution which corresponds to $X + Y$. We also want to obtain a solution 3.2% in ethanol, so the amount of ethanol should be $0.032 \times 60 = 1.92$. The total amount of ethanol is given by the amount in $X(0.02X)$ plus the amount in $Y(0.06Y)$. Translating into algebraic expressions:

$$\begin{cases} X + Y = 60 \\ 0.02X + 0.06Y = 1.92 \end{cases}$$

By solving the system above we obtain:

$$X = 42 \text{ and } Y = 18$$

29. **(D)**

For $x = 2$, the equation becomes

$$\sum_{k=0}^{5} (8 + k) = P + \sum_{k=0}^{5} k.$$

This can be factored to

$$\sum_{k=0}^{5} 8 + \sum_{k=0}^{5} k = P + \sum_{k=0}^{5} k.$$

Therefore,

$$P = \sum_{k=0}^{5} 8 = (6)(8) = 48.$$

30. **(A)**

$$3 \arctan \frac{1}{\sqrt{3}} = \arcsin x.$$

Let $\quad \theta = \arctan \frac{1}{\sqrt{3}}$

Then, $\quad \tan\theta = \frac{1}{\sqrt{3}}.$

We can construct a right triangle with

$$\tan\theta = \frac{1}{\sqrt{3}}.$$

From the Pythagorean Theorem, the length of the hypotenuse is

$$\sqrt{(\sqrt{3})^2 + 1^2} = 2$$

These are the sides of a 30°-60°-90° triangle. Hence

$$\theta = 30° = \frac{\pi}{6} \text{ and } \arcsin x = 3\left(\frac{\pi}{6}\right) = \frac{\pi}{2}$$

This implies $x = \sin^{\pi}/_2 = 1$.

A more general method would be evaluating

$$\sin\left(\arctan\frac{1}{\sqrt{3}}\right)$$

and then evaluating x by using the formula:

$$\sin 3\theta = 3\sin\theta - 4\sin^3\theta$$

To evaluate

$$\sin\left(\arctan\frac{1}{\sqrt{3}}\right),$$

first we can find

$$\sec\left(\arctan\frac{1}{\sqrt{3}}\right)$$

by writing

$$1 + \tan^2\theta = \sec^2\theta$$

and then by using the formula

$$\sin^2\theta = 1 - \frac{1}{\sec^2\theta},$$

we will evaluate

$$\sin\left(\arctan\frac{1}{\sqrt{3}}\right).$$

31.   **(A)**

$$\log_3\left(\frac{1}{27}\right) = \log_3\left(\frac{1}{3^3}\right) = \log_3(3^{-3}) = -3.$$

We could also use the identity:

$$\log_a b^c = c\log_a b$$

and write:

$$\log_3\frac{1}{27} = \log_3 3^{-3} = -3\log_3 3 = -3$$

32.   **(C)**

Since $g(x) = 2$ for all x, $g(f(1)) = 2$.

33. **(E)**

$$\sin\left(x + \frac{\pi}{4}\right) = \sin x \cos\frac{\pi}{4} + \cos x \sin\frac{\pi}{4}$$ (I)

Since $\sin x = \frac{1}{2}$, $\cos x = \sqrt{1 - \frac{1}{4}} = \frac{\sqrt{3}}{2}$

Note we only use the positive sign because $0 \le x \le \pi/2$.
   Substituting into (I), we obtain:

$$\frac{1}{2} \times \frac{\sqrt{2}}{2} + \frac{\sqrt{3}}{2} \times \frac{\sqrt{2}}{2}$$

$$\frac{\sqrt{2}}{4} + \frac{\sqrt{3} \times \sqrt{2}}{4} = \frac{\sqrt{2}}{4}(1 + \sqrt{3})$$

34. **(A)**

$$(\sin\theta \times \cot\theta)^2 + (\cos\theta \tan\theta)^2 = \left(\sin\theta\left(\frac{\cos\theta}{\sin\theta}\right)\right)^2 + \left(\cos\theta\left(\frac{\sin\theta}{\cos\theta}\right)\right)^2$$

$$= \cos^2\theta + \sin^2\theta = 1$$

35. **(A)**

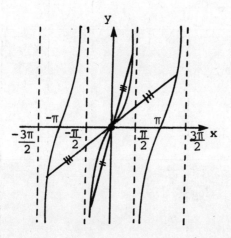

We can translate the statement into the equation

$$yx^2 = k,$$

where k is a constant. This equation indicates that when y is changed, x will change accordingly so that the product of y and x^2 will remain constant. Hence, if y is divided by 4, a 4 will be needed in the numerator. This will happen if x is multiplied by 2; squaring it will supply the 4 in the numerator.

36.   **(A)**

$$\log_3\left(\frac{1}{81}\right) = \log_3\left(\frac{1}{3^4}\right) = \log_3(3^{-4}) = -4.$$

37.   **(B)**
First, find the radius of the circle.

$$x^2 + 4x + y^2 + 4y = x^2 + 4x + 4 + y^2 + 4y + 4 - 4 - 4 = 0$$

We then have an explicit equation for a circle:

$$(x - 2)^2 + (y - 2)^2 = 8.$$

The radius of the circle is $2\sqrt{2}$. From the figure, we can see that the center of the circle overlaps with the centroid of the triangle. Thus, the center of the circle cuts the height of the triangle into a two-third section and a one-third section of its length. So, we can find the height of the triangle by

$$h = \frac{3}{2} \times 2\sqrt{2} = 3\sqrt{2}$$

Since the triangle is equilateral, AB, whose length is the height h, divides the angle CAD evenly, thus angle CAB is 30°. Therefore, the length of the side of the triangle can be found as

$$\frac{3\sqrt{2}}{\cos 30°} = 4.899$$

38.   **(D)**

$$\frac{\sqrt{-20}}{\sqrt{-5}} = \frac{i\sqrt{20}}{i\sqrt{5}} = \frac{\sqrt{4}\sqrt{5}}{\sqrt{5}} = \sqrt{4} = 2.$$

39.   **(C)**
Since we are performing the operation on the given system with $x = x_1 - 4$ and $y = y_1 + 2$, we have

$$2(x_1 - 4) + 3(y_1 + 2) - 18 = 0$$

$$2x_1 - 8 + 3y_1 + 6 - 18 = 0$$

Hence $2x_1 + 3y_1 - 20 = 0$ is the evaluation of 1 in the above system.

40.   **(B)**

   Distance = (rate) × (time).

Let x = time (hours) for the second car to overtake the first. The distance of the second car in x hours is $60x$. The distance of the first car in x hours is $40x$. Since the second car has to travel 5 more miles we have

$$40x + 5 = 60x$$

or $20x = 5 \Rightarrow x = \dfrac{1}{4}$ hr.

41.   **(C)**

   This is a typical second order equation, so the formula

$$r_{1,2} = \frac{-b \pm \sqrt{b^2 - 4ac}}{2a}$$

can be used. The two roots are 2.87 and –0.87.

42.   **(C)**

$$4(2x - 6) - 10x \leq -28$$

$$8x - 24 - 10x \leq -28$$

$$-2x - 24 \leq -28$$

$$-2x \leq -4$$

or $$\frac{-2x}{-2} \geq \frac{-4}{-2}$$

Hence, $x \geq 2$.

43.   **(B)**

   The roots of the equation are 2, –3, and $^7/_5$. Hence, $x = 2$, $x = -3$, and $x = {}^7/_5$. Hence,

$$(x - 2)\,(x + 3)\,(x - \frac{7}{5}) = 0.$$

Multiply both sides of this equation by 5:

$$5(x - 2)\,(x + 3)\,(x - \frac{7}{5}) = 5(0)$$

or $$(x - 2)\,(x + 3)\,5(x - \frac{7}{5}) = 0$$

or
$$(x - 2)(x + 3)(5x - 7) = 0$$

$$(x^2 + x - 6)(5x - 7) = 0$$

$$5x^3 - 7x^2 + 5x^2 - 7x - 30x + 42 = 0$$

$$5x^3 - 2x^2 - 37x + 42 = 0.$$

44. **(C)**
The distance d of a point (x_0, y_0) to the line

$$Ax + By + C = 0$$

is $d = \dfrac{|Ax_0 + By_0 + C|}{\sqrt{A^2 + B^2}}$

For the problem given here, we therefore have

$$d = \dfrac{|2 \times 2 + 1 \times 4 - 1|}{\sqrt{2^2 + 1^2}} = \sqrt{5} = 2.236$$

45. **(B)**
$\cos 35° = 0.819$. Then from the equation,

$$\tan x° = 0.819.$$

So, $x = \tan^{-1}(0.819) = 39.32°$.

46. **(B)**

$$(0.0081) = (0.3)^4$$

Hence

$$(0.0081)^{-\frac{3}{4}} = (0.3^4)^{-\frac{3}{4}} = (0.3)^{-3}$$

$$= \dfrac{1}{0.3^3} = \dfrac{1}{0.027} = \dfrac{1}{\dfrac{27}{1,000}} = \dfrac{1,000}{27}$$

47. **(D)**
Use the surface area formula

$$S = 4\pi r^2$$

and the volume formula

$$V = \frac{4}{3}\pi r^3.$$

From the given condition that $S = 10$:

$$r = \sqrt{\frac{10}{4\pi}}.$$

Thus, its volume is

$$V = \frac{4}{3}\pi \sqrt{\left(\frac{10}{4\pi}\right)^3} = 2.97$$

48.   **(D)**
   With two dice there are 36 possible outcomes. For six of these rolls the dice add up to 7. Thus, the probability that 7 will show on a single toss is $P = \frac{1}{6}$. The probability that 7 will not show is $1 - p = \frac{5}{6}$. Hence the probability that 7 will show on the first 3 tosses and not on the other two is

$$\frac{1}{6} \times \frac{1}{6} \times \frac{1}{6} \times \frac{5}{6} \times \frac{5}{6} = \left(\frac{1}{6}\right)^3 \left(\frac{5}{6}\right)^2$$

49.   **(C)**

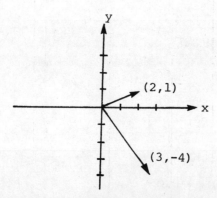

When adding vectors we add their respective components. First we must express the vectors as the distance from the origin to the given point. The point $M(2, 1)$ is the same as the vector **OM**. So we can represent the vector **MN** as the sum of the distance from the origin to M plus the distance from the origin to N. Thus,

$$\mathbf{MN} = \mathbf{MO} + \mathbf{ON}$$

$$= -\mathbf{MO} + \mathbf{ON}$$

$$= -(2, 1) + (3, -4)$$

$$= (1, -5)$$

Thus the distance from M to N is the same as the distance from the origin to $(1, -5)$. Similarly,

$$\mathbf{AB} = -\mathbf{OA} + \mathbf{OB}$$

$$= -(1, -1) + (2, 3)$$

$$= (1, 4)$$

$$\mathbf{CD} = -(-4, 5) + (-3, 10)$$

$$= (1, 5)$$

$$\mathbf{EF} = -(3, -2) + (4, -7)$$

$$= (1, -5)$$

$$\mathbf{GH} = -(3, 2) + (7, -4)$$

$$= (5, -6)$$

$$\mathbf{KL} = -(-2, 3) + (-7, 4)$$

$$= (-5, 1)$$

Hence the vector $\mathbf{MN}$ is equivalent to the vector $\mathbf{EF}$.

50.　　**(B)**
    The area of quadrilateral $ABCD$ is the sum of the areas of $\triangle AOD$ and the trapezoid $OBCD$.

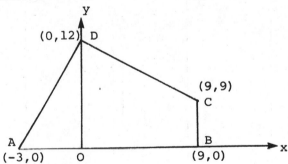

Area of triangle $= \dfrac{1}{2}bh = \dfrac{1}{2}(3)(12) = 18$

Area of trapezoid $= \dfrac{h}{2}(b + b')$

where b and b' are the bases, and h is the altitude. Therefore,

Area of $OBCD = \dfrac{9}{2}(12 + 9) = \dfrac{9}{2}(21) = 94.5$

Total area $= 18 + 94.5 = 112.5$.

THE SAT SUBJECT TEST IN

Math
Level 2

PRACTICE TEST 3

SAT Mathematics Level 2

Practice Test 3

Time: 1 Hour
       50 Questions

DIRECTIONS: Choose the best answer for each question and mark the letter of your selection on the corresponding answer sheet in the back of the book.

NOTES:

(1) Some questions require the use of a calculator. You must decide when the use of your calculator will be helpful.

(2) You may need to decide which mode your calculator should be in—radian or degree.

(3) All figures are drawn to scale and lie in a plane unless otherwise stated.

(4) The domain of any function f is the set of all real numbers x for which $f(x)$ is a real number, unless other information is provided.

REFERENCE INFORMATION: The following information may be helpful in answering some of the questions.

Volume of a right circular cone with radius r and height h $V = \dfrac{1}{3}\pi r^2 h$

Lateral area of a right circular cone with circumference of the c and slant height l $S = \dfrac{1}{2}cl$

Volume of a sphere with radius r $V = \dfrac{4}{3}\pi r^3$

Surface Area of a sphere with r $S = 4\pi r^2$

Volume of a pyramid with base area B and height h $V = \dfrac{1}{3}Bh$

1. The vertices of $\triangle ABC$ are $A(-3, 0)$, $B(3, 0)$, and $C(0, 2)$. The triangle ABC is therefore

    (A) equilateral.

    (B) isosceles.

    (C) scalene.

    (D) right angular.

    (E) indeterminate, due to insufficient information.

2. From a certain point on a level plain at the foot of a mountain, the angle of elevation of the peak is 45°. From a point 50 feet farther away, the angle of elevation of the peak is 30°. What is the height of the peak above the foot of the mountain?

    (A) $\dfrac{50}{\sqrt{3}}$ (D) $50\sqrt{3}$

    (B) $50(\sqrt{3} - 1)$ (E) $50\sqrt{3} + 1$

    (C) $\dfrac{50}{\sqrt{3} - 1}$

3. In the function

    $$f(x) = \frac{2x^2 + 3x + 5}{x^2 - 5x + 5},$$

    x cannot be

    (A) 3.62 and 1.38   (D) 3.15 and 6.24

    (B) 0.25 and 1.37   (E) 10.2 and 2.3

    (C) 5.2 and −7.8

4. A bag contains 6 white balls, 3 red balls, and one blue ball. If one ball is drawn from the bag, what is the probability it will be white?

    (A) 0.1   (D) 0.6

    (B) 0.3   (E) $0.6\overline{6}$

    (C) $0.3\overline{3}$

5. Express cos(arctan$x$) without trigonometric functions.

(A) $\sqrt{1+x^2}$

(D) $\dfrac{x}{\sqrt{1-x^2}}$

(B) $\dfrac{1}{\sqrt{1+x^2}}$

(E) $\dfrac{1}{\sqrt{1-x^2}}$

(C) $\dfrac{1}{x}$

6. The solution set of $5^{2x^2} \times 5^{4x} = 25^{-1}$ is:

(A) {1}

(D) { }

(B) {−1}

(E) $\sqrt{-1}$

(C) {1, −1}

7. A right triangle is shown in the figure, where $h = 5$ and angle $A = 20°$. Find the area of this triangle.

(A) 20.24

(B) 18.57

(C) 38.89

(D) 30.96

(E) 21.35

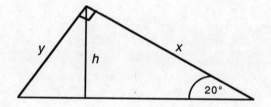

8. The crossing point between the two lines,

$x + 2y - 4 = 0$ and $-2x + y + 3 = 0$,

is d distance away from the origin. Find d.

(A) 0.754

(D) 2.236

(B) 6.453

(E) 3.234

(C) 1.885

9. If $f(x) = 2x^2 + 4x + k$, what should the value of k be in order to have the graph of this function intersect the x-axis in only one place?

(A) –2   (D) 2

(B) 4   (E) –4

(C) 1

10. If a right triangle is rotated 360° around one of its legs as an axis, the solid generated is a

(A) cube.   (D) ellipsoid.

(B) cone.   (E) cylinder.

(C) sphere.

11. In a class of 20 students, a student speaks German, Chinese, or both. Thirteen students speak German and 17 speak Chinese. How many speak both German and Chinese?

(A) 5   (D) 15

(B) 10   (E) 17

(C) 13

12. $t = -9$ is a root of the equation $t^2 + 4t - 45$. Which of the following statements is(are) correct for the equation?

I. $t - 9$ is a factor of the equation.

II. Division of the equation by $t - 9$ yields the other factor of the quadratic equation.

III. $t = -5$ is another root of the equation.

(A) I only.

(B) II and III only.

(C) III only.

(D) I, II, and III.

(E) None of the statements are correct.

13. The function

$$f(x) = 1 + \sqrt{\sin^2(x) + 1}$$

has the maximum value of

(A) 2

(D) 1.41

(B) 3

(E) 2.41

(C) 1

14. The inverse of the function

$$y = \log_2 \frac{2x - 1}{2} \text{ is:}$$

(A) $y = \dfrac{4^x + 1}{2}$

(D) $y = 2^x$

(B) $y = \dfrac{2^x + 1}{2}$

(E) $y = \dfrac{2^{x+1} + 1}{2}$

(C) $y = \dfrac{2^{x+1}}{2}$

15. If $f(x) = x^2 - x - 3$,

$$g(x) = \frac{(x^2 - 1)}{(x + 2)}$$

and $\quad h(x) = f(x) + g(x)$,

find $h(2)$.

(A) 0

(D) $-\dfrac{1}{4}$

(B) $\dfrac{1}{2}$

(E) $\dfrac{7}{4}$

(C) $-\dfrac{1}{2}$

16. $$\dfrac{\sin\dfrac{\pi}{3} + \cos\dfrac{2\pi}{3}}{\tan\dfrac{7\pi}{4}} =$$

(A) $\dfrac{1}{2} - \sqrt{3}$

(D) $\dfrac{\sqrt{3} - 1}{2}$

(B) $\sqrt{3}$

(E) $\dfrac{1 - \sqrt{3}}{2}$

(C) $1 - \sqrt{3}$

17. Which of the graphs below represents the function

$$y = 2 + \sin(x - \pi)?$$

(A)

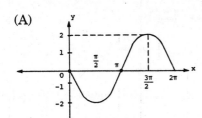

(D)

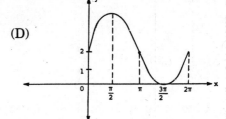

(B)

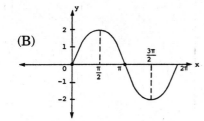

(E)

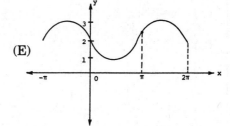

(C)

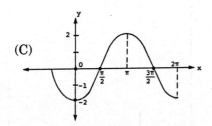

18. If $f(x) = 2^x + 4$ and $g(x) = \dfrac{1}{x}$, then $g(f(g(x)))$ is:

(A) $\dfrac{1}{2^x + 4}$ (D) $\dfrac{1}{\dfrac{1}{2^x} + 4}$

(B) $2^{\frac{1}{x}} + 4$ (E) $\dfrac{1}{2^{2x}} + 4$

(C) $\dfrac{1}{2^{x+4}}$

19. A cube is inscribed in a sphere. If the sphere has a volume of 24, what is the length of the diagonal of the cube?

(A) 3.58   (D) 8.32

(B) 4.27   (E) 11.2

(C) 1.59

20. Obtain the equation of the line passing through point $P(-1, -3)$ and with an inclination of 45°.

(A) $y = x + 2$ (D) $y = x + 1$

(B) $y = x + 4$ (E) $y = x - 2$

(C) $y = -x + 4$

21. If $2\arcsin \dfrac{1}{\sqrt{2}} = \arccos x$, then x equals

(A) $-\dfrac{1}{2}$ (D) $\dfrac{1}{\sqrt{2}}$

(B) 0   (E) $\dfrac{1}{\sqrt{3}}$

(C) $\dfrac{1}{2}$

22. What is $2x^2 + 5x - 12$ divided by $x + 4$?

(A) $x^2 + 2x - 1$ (B) $2x^2 - 8x + 1$

(C) $2x + 3$ (D) $2x - 8$

(E) $2x - 3$

23.   Determine $\lim\limits_{x \to \infty} \dfrac{5x^3 + x^2 - 3x + 5}{x^3 - 1}$.

(A)   5   (D)   2

(B)   –3   (E)   4

(C)   1

24.   What is the domain of the function $\log(5x^2 - 7)$?

(A) $x^2 < 1.4$ (D) $x > -1.18$

(B) $x^2 > 1.4$ (E) $x < 1.18$

(C) $| x | > 1.183$

25.   Simplify the expression $\dfrac{4! + 3!}{5!}$

(A) $\dfrac{3}{10}$ (D) $\dfrac{3}{20}$

(B) $\dfrac{1}{5}$ (E) $\dfrac{12}{5}$

(C) $\dfrac{1}{4}$

26.   In the interval $0 \le x \le \pi$, what is x when $\sin(10x)$ passes the x-axis the third time?

(A)   23.6   (D)   0.324

(B)   62.8   (E)   0.628

(C)   1.325

27. If $f(x) = 2x^2 - 3x + 4$, what value of x will make $f(x)$ a minimum?

(A) $\dfrac{3}{4}$

(D) -1

(B) $\dfrac{3}{2}$

(E) -2

(C) 0

28. The expression

$$\frac{1}{1 - \cos A} + \frac{2}{1 + \cos A}$$

is equivalent to:

(A) $\dfrac{3 \cos A - 1}{\sin^2 A}$

(D) $\dfrac{3 + 3 \cos A}{\sin^2 A}$

(B) $\dfrac{-1 - \cos A}{\sin^2 A}$

(E) $\dfrac{3}{1 - \cos A}$

(C) $\dfrac{3 - \cos A}{1 - \cos^2 A}$

29. Determine the area of triangle ABC if:

I. $A \equiv (1, -1)$ and $B \equiv (10, -1)$

II. $y = 2x - 3$ is the equation of the side $\overline{AC}$.

III. The slope of line $\overline{BC}$ is infinite.

(A) 100

(D) 90

(B) 121

(E) 81

(C) 76.5

30. Let $f(x) = x^2 - x - 6$. Where does the graph of the function cross the x-axis?

(A) $\{(0, -6)\}$

(B) $\{(0, -3), (0, 2)\}$

(C) $\{(-3, 0), (2, 0)\}$

(D) $\{(3, 0), (-2, 0)\}$

(E) The graph does not cross the x-axis.

31. In a school of 300 students, 260 students take physics and 185 students take calculus. How many students take both physics and calculus if we know that every student takes at least one of the courses?

(A) 125

(D) 145

(B) 130

(E) 150

(C) 135

32. $f(a, b) = a^4 + a^2b^2 - a^2$ for all real numbers a and b. Indicate which of the following is true.

I. $f(a, b) = f(a, -b)$

II. $f(a, b) = f(-a, b)$

III. $f(a, b) = f(-a, -b)$

(A) I only.

(D) I and III only.

(B) II only.

(E) I, II, and III.

(C) III only.

33. $f(x) = x^3 + 2x - 1$, then $f(2 - a) =$

(A) $8 - 8a + 2a^2 - 4a - 1$

(B) $8 - 8a - 4a + a^3 - 1$

(C) $11 - 14a + 6a^2 - a^3$

(D) $14a + 8 - 6a^2 + a^3$

(E) Cannot be determined.

34. What is the solution set of the inequality $\dfrac{4}{x-2} < 2$?

(A) $\{x > 4\}$

(D) $\{x < 2\} \cup \{x > 4\}$

(B) $\{x < 4\}$

(E) ϕ

(C) $\{x < 4\} \cup \{x > 4\}$

35. Find all values of x for which $x^{\log_{10} x} = 100x$.

(A) $\{1, {}^1/_{10}\}$

(D) $\{10, 100\}$

(B) $\{{}^1/_{10}, 10\}$

(E) $\{{}^1/_{10}, 100\}$

(C) $\{100, 1000\}$

36. 10 white balls and 19 blue balls are in a box. If a man draws a ball from the box at random, what are the odds in favor of him drawing a blue ball?

(A) 10:29

(D) $\dfrac{10}{19} : \dfrac{19}{29}$

(B) 19:29

(E) $\dfrac{19}{10} : \dfrac{29}{10}$

(C) 19:10

37. If $27^x = 9$ and $2^{x-y} = 64$, then $y =$

(A) -5

(D) $-\dfrac{11}{3}$

(B) -3

(E) $-\dfrac{16}{3}$

(C) $-\dfrac{2}{3}$

38. The fourth term of $(1 - 2x^2)^9$ is:

(A) $24x^3$

(D) $4x^6$

(B) $-223x^5$

(E) $-672x^6$

(C) $1249x^6$

39. If $x^{-2} - 9 = 0$, solve for x.

(A) 3

(D) ± 3

(B) $-\dfrac{1}{3}$

(E) $\pm \dfrac{1}{3}$

(C) $\dfrac{1}{3}$

40. If $\cos x = \dfrac{4}{5}$ and $0 \le x \le \dfrac{\pi}{2}$, then $\tan 2x =$

    (A) $\dfrac{7}{24}$ (D) $\dfrac{7}{25}$

    (B) $\dfrac{3}{5}$ (E) $\dfrac{24}{7}$

    (C) $\dfrac{24}{25}$

41. If $x = 1$ and $f(x) = \sqrt{x} + 1$, then $f(f(f(x)))$ equals

    (A) 4.35   (D) 1.32

    (B) 1.75   (E) 2.55

    (C) 3.24

42. Given that $\arccos \dfrac{4}{5}$ is in the first quadrant, find $\sin(\arccos \dfrac{4}{5})$.

    (A) $\dfrac{1}{2}$ (D) 1

    (B) $\dfrac{2}{5}$ (E) Cannot be determined.

    (C) $\dfrac{3}{5}$

43. Consider the function

    $$f(x) = x^3 + 3x - k.$$

    If $f(2) = 10$, then $k =$

    (A) 4   (D) 18

    (B) -4 (E) 14

    (C) 0

44. 72° corresponds to an angle measure of

    (A) $\dfrac{2\pi}{5}$ (B) $\dfrac{\pi}{5}$

(C) 2π (D) $\dfrac{\pi}{4}$

(E)   Cannot be determined.

45.   Find the smallest positive value of x so that $\cos(x + 985°) = 1$.

(A)   455   (D)   95

(B)   635   (E)   275

(C)   0

46.   Express $\log_b 2 + \log_b \beta + \dfrac{1}{2}\log_b \gamma - \dfrac{1}{2}\log_b \omega$ as a single logarithm.

(A) $\dfrac{1}{2}\log_b\left(2\beta\dfrac{\gamma}{\omega}\right)$ (D) $\dfrac{1}{2}\log_b\left(\beta\sqrt{\dfrac{\omega^2}{\gamma^2}}\right)$

(B) $2\log_b(2\beta\omega\gamma)$ (E) $2\log_b\left(\beta\sqrt{\dfrac{\omega}{\gamma}}\right)$

(C) $\log_b\left(2\beta\sqrt{\dfrac{\gamma}{\omega}}\right)$

47.   If $x_0 = 1$, $x_1 = 2$ and

$$\dfrac{x_{n+1}}{x_n} = \sqrt{\dfrac{x_n}{x_{n-1}} + 2}, x_4 \text{ is}$$

(A)   16   (D)   28

(B)   8   (E)   12

(C)   24

48.   The base of a right prism, shown here, is an equilateral triangle, each of whose sides measure 4 units. The altitude of the prism is 5 units. Find the volume of the prism.

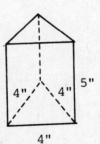

4"   4"   5"

4"

(A) $4\sqrt{3}$

(D) 40

(B) $20\sqrt{3}$

(E) $15\sqrt{4}$

(C) 60

49. Find the value of e in $\ln(e) = 1$.

(A) 1.414

(D) 2.718

(B) 3.713

(E) 3.14

(C) 10

50. If $(3x)^{\frac{3}{2}} + (\frac{8}{3}x)^{\frac{2}{3}} = 31$, what should x be?

(A) 1

(D) 3

(B) 10

(E) 7

(C) 6

SAT Mathematics
Level 2

Practice Test 3
ANSWER KEY

1.	(B)	14.	(E)	27.	(A)	40.	(E)
2.	(C)	15.	(D)	28.	(C)	41.	(E)
3.	(A)	16.	(E)	29.	(E)	42.	(C)
4.	(D)	17.	(E)	30.	(D)	43.	(A)
5.	(B)	18.	(D)	31.	(D)	44.	(A)
6.	(B)	19.	(A)	32.	(E)	45.	(D)
7.	(C)	20.	(E)	33.	(C)	46.	(C)
8.	(D)	21.	(B)	34.	(D)	47.	(A)
9.	(D)	22.	(E)	35.	(E)	48.	(B)
10.	(B)	23.	(A)	36.	(C)	49.	(D)
11.	(B)	24.	(C)	37.	(E)	50.	(D)
12.	(E)	25.	(C)	38.	(E)		
13.	(E)	26.	(E)	39.	(E)		

DETAILED EXPLANATIONS OF ANSWERS

1. **(B)**

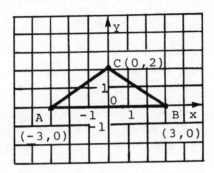

If we let the origin be point O, and prove $\triangle COA \cong \triangle COB$, then, by corresponding parts, we can conclude $\overline{CA} \cong \overline{CB}$. This will be sufficient to show $\triangle ABC$ is isosceles, since an isosceles triangle is defined to be one in which two sides are congruent. The SAS $\cong$ SAS method will be used.

Since, by definition of the Cartesian plane, the y-axis $\perp x$-axis; thus angle COA and angle COB are right angles and, they are congruent.

$OA = 3$ units and $OB = 3$ units and, because their lengths are equal, they are congruent.

We now have congruence between one angle in each triangle and one corresponding adjacent side. The other adjacent side, $\overline{OC}$, is common to both triangles and, by reflexivity of congruence, is congruent to itself.

Therefore, $\triangle COA \cong \triangle COA$ by SAS $\cong$ SAS.

Thus, $\overline{CA} \cong \overline{CB}$, because corresponding sides of congruent triangles are congruent.

Therefore, $\triangle ABC$ is isosceles because it is a triangle which has two congruent sides.

2. **(C)**

The situation is illustrated in the figure. Triangles APC and BPC are right triangles, so

$$\tan 45° = \frac{y}{x}, \tan 30° = \frac{y}{x+50}$$

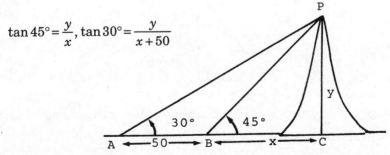

Since tan 45° = 1 and tan 30° = $1/\sqrt{3}$, these equations become

$$y = x,$$

and $\quad x + 50 = \sqrt{3}\,y.$

When we solve these last two equations for y, we get:

$$y = \frac{50}{\sqrt{3} - 1}$$

3. **(A)**

The domain of this function is constrained to x that does not make $f(x)$ become infinite, in which case we say the function is defined on that domain. The values of x that can make the function undefined are the roots of the denominator,

$$x^2 - 5x + 5.$$

Using the following formula, you can find these roots:

$$r_{1,2} = \frac{-b \pm \sqrt{b^2 - 4ac}}{2a}$$

Since here $a = 1$, $b = -5$, $c = 5$, you can easily find these roots to be 3.62 and 1.38.

4. **(D)**

There are 10 balls in all and 6 of them are white. Hence, the probability of drawing a white ball is

$$\frac{6}{10} = 0.6$$

5. **(B)**

Let $\theta = \arctan x$. Then $\tan\theta = x$. We can construct a right triangle with $\tan\theta = x$:

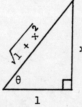

The length of the hypotenuse, obtained by the Pythagorean Theorem, is $\sqrt{1 + x^2}$. Thus $\cos(\arctan x) =$

$$\cos\theta = \frac{1}{\sqrt{1+x^2}}$$

6. **(B)**

The expression

$$5^{2x^2+4x} = 25^{-1}$$

can be rewritten as:

$$5^{2x^2+4x} = (5^2)^{-1} = 5^{-2}$$

Since the base on both sides is the same:

$$2x^2 + 4x = -2$$

If we divide both sides by 2 we obtain:

$$x^2 + 2x = -1, \text{ or } x^2 + 2x + 1 = 0$$

The equation above has a root -1 with multiplicity 2, so the answer is $\{-1\}$.

7. **(C)**

$$x = \frac{5}{\sin(20°)} = 14.62$$

and $y = 14.62 \times \tan 20° = 5.32$.

The area of the triangle is

$$\frac{1}{2}x \times y = \frac{1}{2} \times 14.62 \times 5.32 = 38.89.$$

8. **(D)**

Find the crossing point by treating these two lines with regard to their corresponding equations. Multiplying the first equation by 2 and adding it to the second will give you $5y = 5$, or $y = 1$. Then, plugging this value into either equation will give $x = 2$. So, the point of crossing is (2, 1) and its distance to the origin is

$$\sqrt{(2^2 + 1^1)} = \sqrt{5} = 2.236$$

9. **(D)**

The function given is represented graphically by a parabola. A parabola will intersect the x-axis in only one place if both roots are equal, which means $\Delta = 0$.

$$\Delta = b^2 - 4ac = +(4)^2 - 4(2)k = 0$$

$$16 - 8k = 0$$

$$16 = 8k \quad \therefore k = 2$$

10. **(B)**

The described solid is a cone. The perpendicular cross-section of a cone which contains the central axis is an isosceles triangle and when this triangle is bisected we get two right triangles.

11. **(B)**

(All students) = (German speakers) + (Chinese speakers)

– (Speakers of both languages)

$20 = 13 + 17 - x.$

Hence, $x = 10.$

12. **(E)**

Since $t = -9$ is a root of the equation

$$t^2 + 4t - 45,$$

then $(t + 9)$ is a factor. Division of the equation by $(t + 9)$ would therefore yield the other factor of the quadratic, which is $(t - 5)$, giving the second root, $t = 5.$

13. **(E)**

The $\sin(x)$ function is bounded between 1 and –1. So, $\sin^2(x)$ has a maximum value of 1. Since the square root function is an increasing function, thus the function $f(x)$ has its maximum value at

$$1 + \sqrt{1 + 1} = 2.41.$$

14. **(E)**

To determine the inverse of a function it is necessary to replace the variable y by x and x by y.

The function

$$y = \log_2 \frac{2x - 1}{2}$$

becomes:

$$x = \log_2 \frac{2y - 1}{2}$$

$$2^x = \frac{2y - 1}{2}$$

$$\frac{2^{x+1} + 1}{2} = y$$

15.　**(D)**

$$h(x) = f(x) + g(x),$$

and we are told that

$$f(x) = x^2 - x - 3 \text{ and } g(x) = (x^2 - 1)\,(x + 2);$$

thus　　$h(x) = (x^2 - x - 3) + \dfrac{x^2 - 1}{x + 2}.$

To find $h(2)$, we replace x by 2 in the above formula for $h(x)$:

$$h(2) = [(2)^2 - 2 - 3] + \left(\frac{2^2 - 1}{2 + 2}\right)$$

$$= (4 - 2 - 3) + \left(\frac{4 - 1}{4}\right)$$

$$= (-1) + \left(\frac{3}{4}\right)$$

$$= -\frac{4}{4} + \frac{3}{4}$$

$$= -\frac{1}{4}$$

Thus, $h(2) = -\dfrac{1}{4}.$

16.　**(E)**
　　Evaluate each term in the numerator and denominator separately:

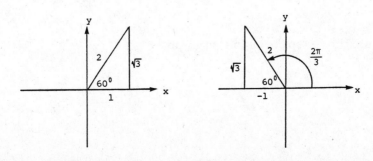

$$\sin\frac{\pi}{3} = \sin 60° = \frac{\sqrt{3}}{2}, \quad \cos\frac{2\pi}{3} = -\frac{1}{2}$$

$$\frac{\sin\dfrac{\pi}{3} + \cos\dfrac{2\pi}{3}}{\tan\dfrac{7\pi}{4}} = \frac{\dfrac{\sqrt{3}}{2} + \left(-\dfrac{1}{2}\right)}{-1}$$

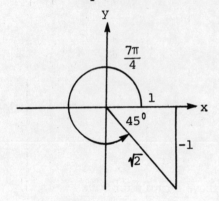

$$\tan\frac{7\pi}{4} = \frac{-1}{1} = -1,$$

$$= \frac{\sqrt{3} - 1}{-2}$$

$$= \frac{1 - \sqrt{3}}{2}$$

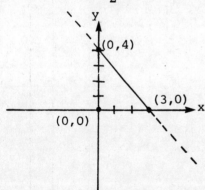

17.　**(E)**
There are three steps to build this graph.

First Step: Draw the graph $y = \sin x$

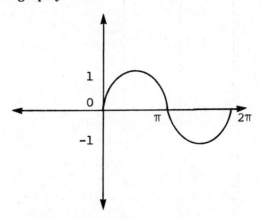

Second Step: Move the graph two units up, because it is $y = \sin(x - \pi) + 2$.

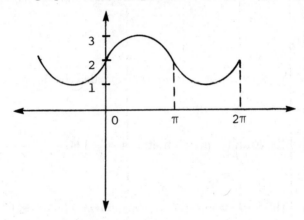

Third Step: Move the graph π units because it is $y = \sin(x - \pi) + 2$.

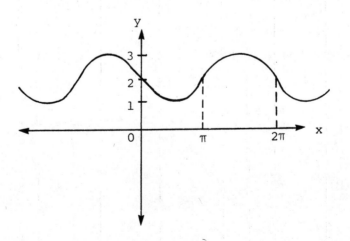

18. **(D)**

$$g(f(g(x)))$$

$$= g(f(\frac{1}{x}))$$

$$= g(2^{\frac{1}{x}} + 4)$$

$$= \frac{1}{2^{\frac{1}{x}} + 4}$$

19. **(A)**

The diagonal of the cube is also the diameter of the sphere. By using the formula

$$V = \frac{4}{3}\pi r^3,$$

we can find the radius of the sphere, thus also the diameter. Specifically,

$$r^3 = \frac{24 \times 3}{4 \times \pi} = \frac{18}{\pi} = 5.73$$

Thus, $r = 1.79$. Finally, the diameter is $2r = 3.58$.

20. **(E)**

A line with inclination of 45° has a slope m which is equal to tan 45°.
The equation of a line can be written as below:

$$y - y_0 = m(x - x_0)$$

The ordered pair (x_0, y_0) is given and $m = \tan 45° = 1$.

By substitution we obtain:

$$(y - (-3)) = 1(x - (-1))$$

$$y + 3 = x + 1$$

$$y = x - 2$$

21. **(B)**

$$2 \arcsin \frac{1}{\sqrt{2}} = \arccos x,$$

$$2\left(\frac{\pi}{4}\right) = \arccos x,$$

$$x = \cos\left(\frac{\pi}{2}\right) = 0.$$

22.   (E)
If one notices that the given polynomial can be factored into $(x + 4)(2x - 3)$ we have the answer immediately.

Otherwise we can use synthetic division:

$$
\begin{array}{r}
2x - 3 \\
x + 4 \overline{\smash{\big)}\ 2x^2 + 5x - 12} \\
\underline{-(2x^2 + 8x)} \\
-3x - 12 \\
\underline{-(-3x - 12)} \\
0
\end{array}
$$

23.   (A)
By inspection it is possible to rearrange the expression as shown below:

$$\lim_{x \to \infty} \frac{5(x^3 - 1) + x^2 - 3x}{x^3 - 1} = \frac{5(x^3 - 1)}{x^3 - 1} + \frac{x^2 - 3x}{x^3 - 1}$$

$$= \lim_{x \to \infty} 5 + \frac{x^2}{x^3 - 1} - \frac{3x}{x^3 - 1}$$

$$= \lim_{x \to \infty} 5 + \frac{1}{x - \dfrac{1}{x^2}} - \frac{3}{x^2 - \dfrac{1}{x}} = 5$$

24.   (C)
The domain of a $\log(x)$ function is that the variable x must be greater than 0. From this, we have $5x^2 - 7 > 0$, which can be further simplified as

$$x^2 > \frac{7}{5}$$

But this is not yet in the final form. Since $^7/_5 = 1.4$, we have $-1.18 < x < 1.18$.

25.   (C)

$$\frac{4! + 3!}{5!} = \frac{(4)3! + 3!}{(5)(4)\,3!} = \frac{(4 + 1)3!}{(5)(4)\,3!}$$

$$= \frac{4 + 1}{(5)\,4} = \frac{5}{20} = \frac{1}{4}$$

26. (E)

Within the given interval, $\sin(10x)$ passes the x-axis at

$$0, \frac{\pi}{10}, \frac{\pi}{5}, \frac{3\pi}{10}, \text{etc.}$$

Therefore, at the third passing, $x = \pi/5 = 0.628$.

27. (A)

$$f(x) = 2(x^2 - \frac{3}{2}x + 2)$$

$$= 2[(x - \frac{3}{4})^2 - \frac{9}{16} + 2] \quad \text{(completing the squares)}$$

$$= 2[(x - \frac{3}{4})^2 + \frac{23}{16}]$$

Since $(x - 3/4)^2 \geq 0$, the smallest it can be is 0. Hence $x = 3/4$ makes f a minimum.

28. (C)

$$\frac{1}{1 - \cos A} + \frac{2}{1 + \cos A} = \frac{1 + \cos A + 2(1 - \cos A)}{1 - \cos^2 A}$$

$$= \frac{1 + \cos A + 2 - 2\cos A}{\sin^2 A}$$

$$= \frac{-\cos A + 3}{\sin^2 A}$$

29. (E)

To determine the area of a triangle it is necessary to determine the coordinate of the three vertices of the triangle. A and B are given. Point C must have its x-coordinate equal to 10 (since $\overline{BC}$ is a vertical line) and it must be on the line $y = 2x - 3$. This gives a y value of $2(10) - 3 = 17$.

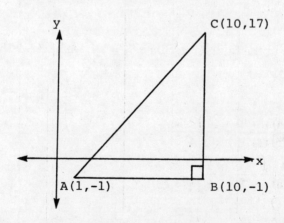

Figure ABC forms a right triangle with base $\overline{AB}$ = 9 and height $\overline{BC}$ = 18.

$$\text{Area}_\triangle = \frac{1}{2} \times \text{base} \times \text{height} = \frac{1}{2} \times 9 \times 18 = 81$$

30. **(D)**
Any point which is located on the x-axis has a y-coordinate of zero. Therefore, to determine where the function crosses the x-axis we must find these values of x for which $y = f(x) = 0$; i.e., we must solve

$$x^2 - x - 6 = 0.$$

We obtain

$$(x - 3)(x + 2) = 0$$

which has solutions $x = 3$ and $x = -2$. The points of intersection are thus $(3, 0)$ and $(-2, 0)$.

31. **(D)**

[all students] = [students taking physics] + [students taking calculus]

– [those taking both].

This formula is a direct conclusion of the following relation:

$$n(A \cup B) = n(A) + n(B) - n(A \cap B)$$

where A and B are two sets and n stands for the number of elements a set has. In the above formula, if we define A = the set of the students who are taking physics and B = the set of the students taking calculus, then $N(A) = 260$ and $n(B) = 185$. We have:

$$n(A \cup B) = 300$$

because each student takes at least one of the two courses. Therefore $(A \cap B)$ will be the set of the students taking both of the courses and $n(A \cap B)$ can be derived from the above relation. The following diagram shows the validity of the equation.

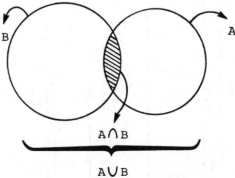

B

A

A∩B

A∪B

The students taking both courses are counted twice in

$$[n(A) + n(B)]$$

and therefore must be subtracted from the sum of students taking individual courses.

We obtain

$$300 = 260 + 185 - x$$

$$= 445 - x$$

Hence, $x = 145$.

32.   **(E)**

It should be evident from the nature of the equation

$$a^4 + a^2b^2 - a^2,$$

that all three choices are correct since any real number taken to an even power would yield a positive real number. Hence

$$f(a, -b) = (a)^4 + a^2(-b)^2 - (a)^2 = a^4 + a^2b^2 - a^2$$

and $f(-a, b) = (-a)^4 + (-a)^2b^2 - (-a)^2$

$$= f(-a, -b) = (-a)^4 + (-a)^2 (-b)^2 - (-a)^2$$

$$= a^4 + a^2b^2 - a^2.$$

33.   **(C)**

$$f(x) = x^3 + 2x - 1$$

$\therefore$ $f(2 - a) = (2 - a)^3 + 2(2 - a) - 1$

$$= [8 - 8a + 2a^2 - 4a + 4a^2 - a^3] + [4 - 2a] - 1$$

$$= 11 - 14a + 6a^2 - a^3.$$

34.   **(D)**

In order for $\dfrac{4}{x-2} < 2$

Either $x - 2 > 2$

therefore $x > 4$

OR if $x - 2 < 0$

then $\dfrac{4}{x-2} < 2$

then $x - 2 < 0$

therefore $x < 2$ also.

35.   **(E)**

Take the log of both sides:

$$\log_{10}(x^{\log_{10}x}) = \log_{10}(100x)$$

$$(\log_{10}x)(\log_{10}x) = \log_{10}(100x)$$

because

$$\log_{10}x^a = a\log_{10}x,$$

setting $a = \log_{10}x$, we will arrive at the above conclusion.

$$(\log_{10}x)^2 = \log_{10}100 + \log_{10}x = 2 + \log_{10}x$$

or $(\log_{10}x)^2 - \log_{10}x - 2 = 0$

This is a quadratic equation and can be factored:

$$(\log_{10}x + 1)(\log_{10}x - 2) = 0$$

$$\log_{10}x = -1 \text{ and } \log_{10}x = 2$$

$\Rightarrow x = 10^{-1} = 1/10$ and $x = 10^2 = 100$.

Note that according to the definition of logarithm, these two statements are equivalent:

$$\log_{10}a = b \Leftrightarrow 10^b = a$$

The solution set is thus $\{1/10, 100\}$.

36.   **(C)**

If an event can happen in p ways and fail to happen in q ways, then, if $p > q$, the odds are p to q in favor of the event happening.

If $p < q$, then the odds are q to p against the event happening. In this case, $p = 19$ and $q = 10$, $p > q$. Thus, the odds in favor of the event of drawing a blue ball are 19:10.

37.   **(E)**

$$27^x = (3^3)^x = 3^{3x}$$

$$9 = 3^2$$

Thus, we have $3^{3x} = 3^2$, which implies $3x = 2$ or $x = 2/3$ because the power function is one-to-one, which means that if

$$a^{x_1} = a^{x_2} \text{ then } x_1 = x_2.$$

Now $\quad 2^{x-y} = 64 = 2^6$

so $\qquad x - y = 6$ or $\dfrac{2}{3} - y = 6.$

Hence, $y = \dfrac{2}{3} - 6 = -\dfrac{16}{3}.$

38.　(E)

A general term is given by the expression:

$$\sum_{r=0}^{n} \frac{n!}{r!(n-r)!} a^{n-r} b^r$$

By substitution we obtain:

$$\frac{9!}{3!6!}(1)^6(-2x^2)^3 = -672x^6$$

39.　(E)

$$x^{-2} - 9 = 0, \; x^{-2} = 9$$

$\therefore \qquad (x^{-2})^{-\frac{1}{2}} = 9^{-\frac{1}{2}}$

$$x = (\pm 3^2)^{-\frac{1}{2}} = \pm 3^{-1}$$

$\therefore \qquad x = \pm \dfrac{1}{3}.$

40.　(E)

If $\cos x = {}^4/_5$, by the trigonometric relationship

$$\sin^2 x + \cos^2 x = 1,$$

we can determine the value of $\sin x$:

$$\sin x = \pm\sqrt{1 - \cos^2 x} = \pm\sqrt{1 - \frac{16}{25}} = \pm\frac{3}{5}$$

Note that only the positive values will be considered since we are working only in the first quadrant.

Since $\sin 2x = 2\sin x \cos x$ and $\cos 2x = \cos^2 x - \sin^2 x$, we obtain:

$$\sin 2x = 2 \times \frac{3}{5} \times \frac{4}{5} = \frac{24}{25}$$

$$\cos 2x = \frac{16}{25} - \frac{9}{25} = \frac{7}{25}$$

Therefore

$$\tan 2x = \frac{\dfrac{24}{25}}{\dfrac{7}{25}} = \frac{24}{7}$$

41.　**(E)**

Notice the function

$$f(f(f(x))) = \sqrt{\sqrt{\sqrt{x+1}+1}+1}$$

Thus you only need to plug $x = 1$ into this equation and find $f(f(f(1)))$, that is

$$f(f(f(1))) = \sqrt{\sqrt{\sqrt{1+1}+1}+1} = \sqrt{\sqrt{\sqrt{2}+1}+1} = 2.55$$

42.　**(C)**

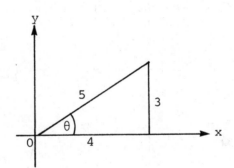

Let

$$\theta = \arccos \frac{4}{5}, \Rightarrow \cos\theta = \frac{4}{5}.$$

We can now construct the above figure. (From the Pythagorean Theorem we know the vertical side is 3).

$$\sin\theta = \frac{3}{5} \Rightarrow \sin(\arccos\frac{4}{5}) = \frac{3}{5}.$$

43.　**(A)**

$$f(x) = x^3 + 3x - k$$

since $f(2) = 10$, we have

$$10 = (2)^3 + 3(2) - k,$$

or $10 = 8 + 6 - 8 \Rightarrow k = 4.$

44.   **(A)**

π radians = 180°

Let x radians = 72°

$\Rightarrow$ $x = \dfrac{72\pi}{180} = \dfrac{2\pi}{5}$ radians.

45.   **(D)**

Let $y = x + 985$. For $\cos(y) = 1$, we have

$$y = 0, \pm 360, \pm 2 \times 360, \ldots$$

But $y = x + 985$, so $x = n \times 360 - 985,$

where $n = 0, \pm 1, \pm 2, \ldots$

So the smallest positive value of x occurs when $n = 3$. Then

$$x = 3 \times 360 - 985 = 95.$$

46.   **(C)**

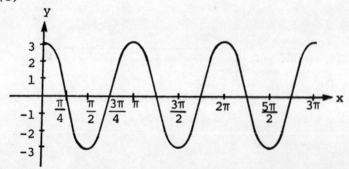

$$\log_b 2 + \log_b \beta + \frac{1}{2}\log_b \gamma - \frac{1}{2}\log_b \omega = (\log_b 2 + \log_b \beta) + \frac{1}{2}(\log_b \gamma - \log_b \omega)$$

$$= \log_b (2\beta) + \frac{1}{2}\log_b\left(\frac{\gamma}{\omega}\right)$$

$$= \log_b\left(2\beta\left(\frac{\gamma}{\omega}\right)^{\frac{1}{2}}\right)$$

$$= \log_b\left(2\beta\sqrt{\frac{\gamma}{\omega}}\right)$$

47. **(A)**

Use the following recursion:

$$x_4 = x_3 \sqrt{\frac{x_3}{x_2} + 2},$$

$$x_3 = x_2 \sqrt{\frac{x_2}{x_1} + 2},$$

$$x_2 = x_1 \sqrt{\frac{x_1}{x_0} + 2}$$

Since $x_0 = 1$ and x_1 are given, you obtain x_2 first. Then by working backward, you can get x_3 and finally x_4. That is,

$$x_2 = 2\sqrt{\frac{2}{1} + 2} = 4,$$

$$x_3 = 4\sqrt{\frac{4}{2} + 2} = 8,$$

$$x_4 = 8\sqrt{\frac{8}{4} + 2} = 16$$

48. **(B)**

We imagine the prism as a stack of equilateral triangles, congruent to the base of the prism. Let each of these triangles be one unit of measure thick. We can then calculate the area of the base, B, and multiply it by the number of bases needed to complete the height of the prism, h, to obtain the volume of the prism. Therefore,

$$V = Bh.$$

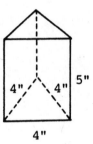

All prism volumes can be thought of in this way.

In this particular problem, the base is an equilateral triangle. Therefore

$$B = \frac{s^2 \sqrt{3}}{4},$$

where s is the length of a side of the base. By substitution,

$$B = \frac{(4)^2 \sqrt{3}}{4} = 4\sqrt{3}.$$

Since the prism is 5 units high,

$$V = Bh = (4\sqrt{3})5 = 20\sqrt{3}.$$

Therefore, the volume of the prism is $20\sqrt{3}$ cu. units.

49.　**(D)**

e is the base of natural logarithm. So, you can find it by directly using your calculator:

$$e = \ln^{-1}1 = 2.71828$$

50.　**(D)**

Plug in each number in the equation and check if the left side is equal to the right side. Only one is correct, which is (D).

THE SAT SUBJECT TEST IN

Math
Level 2

PRACTICE TEST 4

SAT Mathematics Level 2

Practice Test 4

Time: 1 Hour
50 Questions

DIRECTIONS: Choose the best answer for each question and mark the letter of your selection on the corresponding answer sheet in the back of the book.

NOTES:

(1)   Some questions require the use of a calculator. You must decide when the use of your calculator will be helpful.

(2)   You may need to decide which mode your calculator should be in—radian or degree.

(3)   All figures are drawn to scale and lie in a plane unless otherwise stated.

(4)   The domain of any function f is the set of all real numbers x for which $f(x)$ is a real number, unless other information is provided.

REFERENCE INFORMATION: The following information may be helpful in answering some of the questions.

Volume of a right circular cone with radius r and height h	$V = \dfrac{1}{3}\pi r^2 h$
Lateral area of a right circular cone with circumference of the c and slant height l	$S = \dfrac{1}{2}cl$
Volume of a sphere with radius r	$V = \dfrac{4}{3}\pi r^3$
Surface Area of a sphere with r	$S = 4\pi r^2$
Volume of a pyramid with base area B and height h	$V = \dfrac{1}{3}Bh$

1. A cube has total surface area 24 cm². The volume of the cube in cm³ is

   (A)   8   (D)   20

   (B)   12   (E)   36

   (C)   16

2. If a fair die is rolled, what is the probability of obtaining an even number or a number greater than 3?

   (A) $\dfrac{2}{3}$ (D) $\dfrac{5}{6}$

   (B)   1   (E) $\dfrac{1}{6}$

   (C) $\dfrac{1}{3}$

3. If tan$A$ = $^3/_4$ and cos$B$ = $^{12}/_{13}$, where A and B are acute angles, find the value of sin($A + B$).

   (A) $\dfrac{36}{65}$ (D) $\dfrac{16}{65}$

   (B) $\dfrac{20}{65}$ (E)   Cannot be determined.

   (C) $\dfrac{56}{65}$

4. If 13^x = 15, then x^5 =

   (A)   2.5   (D)   1.2

   (B)   1.3   (E)   1.4

   (C)   9

5. Where are the coordinates of the midpoint M of the segment joining the pair of points (1, 2) and (5, 8)?

   (A) $M(6, \dfrac{5}{2})$ (B) $M(3, 5)$

(C) $M(3, \dfrac{5}{2})$ (D) (6, 10)

(E) (6, 5)

6. $f(x) = 5^x + 4$ is a function. If $f(x) = 7$, find x.

(A) 3.32   (D) 2.4

(B) 0.68   (E) 0.5

(C) 0.8

7. The solution set of $2y - x > 6$ lies in which quadrants?

(A) I only.   (D) II and III.

(B) I and II.   (E) I, II, III, and IV.

(C) I, II, and III.

8. Find the surface area of the figure below.

(A) 98

(B) 124

(C) 104

(D) 84

(E) 140

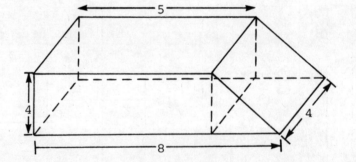

9. Find $\left(\dfrac{27}{35}\right)^{-\frac{2}{3}}$

(A) 0.84   (D) 0.96

(B) 1.19   (E) 1.05

(C) 1.5

10. Find the range, R, of the function

$$f(x) = \frac{x}{|x|}.$$

The domain D, consists of the set of non-zero real numbers.

(A) $R = \{1, 0\}$

(B) $R = \{-1, 1\}$

(C) $(1, -1, 0)$

(D) All positive integers

(E) All non-zero real numbers

11. $\dfrac{n!}{(n+1)!} + \dfrac{(n+1)!}{(n+2)!} =$

(A) $\dfrac{n(2n+3)}{(n+2)}$

(D) $\dfrac{n!}{(N+2)!}$

(B) $\dfrac{2n+3}{(n+1)(n+2)}$

(E) None of the above.

(C) $\dfrac{2n(n+1)!}{(n+2)!}$

12. $\cos(\arcsin\dfrac{1}{2}) =$

(A) $\dfrac{\sqrt{3}}{2}$

(D) $\dfrac{3}{5}$

(B) $\dfrac{1}{2}$

(E) $\dfrac{4}{5}$

(C) $\dfrac{1}{\sqrt{2}}$

13. If $f(x) = \sqrt{x+1}$ and $g(x) = x^3 + 2$, find $f(g(2))$.

(A) 3.32

(D) 10.9

(B) 2.9

(E) 4.2

(C) 7.1

14. What is the remainder when $2x^2 - 4x + 5$ is divided by $x + 1$?

    (A) 9   (D) 15

    (B) 11   (E) 16

    (C) 13

15. There are 6 knights of the roundtable. Given that Sir Lancelot must sit in a specific chair and that Sir Gawain must be directly on either side of him, in how many ways may the knights be seated?

    (A) 24   (D) 25

    (B) 120   (E) 720

    (C) 48

16. Which of the following are odd functions?

    I. $f(x) = 2x + 1$

    II. $g(x) = \tan x$

    III. $h(x) = x^3 + x^2$

    (A) I only.   (D) I, II, and III.

    (B) II only.   (E) None of the above.

    (C) I and II only.

17. $\dfrac{n!}{n(n+1)} =$

    (A) $\dfrac{(n-1)!}{n+1}$ (D) $\dfrac{n}{(n+1)!}$

    (B) $\dfrac{(n-1)!}{(n+1)!}$ (E) $\dfrac{(n-1)!}{n}$

    (C) $\dfrac{n}{n+1}$

18. If $\tan\theta(x + \cot\theta\cos\theta) = \sec\theta$, then $x =$

    (A) $\csc\theta$ (D) 0

    (B) $\cos\theta$ (E) 1

    (C) $\sin\theta$

19. What is $\log_3 6$?

    (A) 0.778   (D) 5.345

    (B) 0.477   (E) 2

    (C) 1.6309

20. If the functions f, g, and h are defined by

    $$f(x) = x^2, \ g(x) = \sin x, \text{ and } h(x) = \frac{x}{2} \text{ then } f(g(h(\pi))) =$$

    (A) 0   (D) $\dfrac{1}{4}$

    (B) 1   (E) $\dfrac{\sqrt{2}}{2}$

    (C) $\dfrac{1}{2}$

21. Point A has coordinates $(2, 5)$ and point B has coordinates $(-3, -3)$. What is the distance between point A and point B?

    (A) 5.6   (D) 2.4

    (B) 9.4   (E) 25.3

    (C) 8.1

22. Express $0.18\overline{18}$ as a fraction.

    (A) $\dfrac{2}{11}$ (D) $\dfrac{3}{17}$

    (B) $\dfrac{2}{13}$ (E) None of the above.

    (C) $\dfrac{1}{9}$

23. Four officers must be chosen from a high school committee of 8 freshmen and 5 sophomores, with exactly 2 officers to be chosen from each class. In how many ways can these officers be chosen?

(A) 1,600

(D) 980

(B) 280

(E) 1,260

(C) 1,680

24. $\lim\limits_{x \to 0}\left(\dfrac{x^2 - 2x}{x}\right) =$

(A) −2

(D) 2

(B) 0

(E) Does not exist.

(C) 1

25. Find the area of the triangle between $f(x) = 2x + 3$ and the x- and y-axes.

(A) 2.25

(D) 6

(B) 4.5

(E) 9

(C) 3.5

26. If $f(x) = x^3$ and $g(x) = x^2 + 1$, which of the following are odd functions?

I. $f(x) \times g(x)$

II. $f(g(x))$

III. $2g(f(x)) - 1$

(A) I only.

(D) I and III only.

(B) II only.

(E) II and III only.

(C) III only.

27. What is the length of the radius of the circle given by the equation

$$x^2 + 4x - 2y + y^2 + 2 = 0?$$

(A) 6   (D) $\sqrt{2}$

(B) 2   (E) $\sqrt{3}$

(C) 3

28. If $\sin x = \dfrac{1}{3}\tan x$, then $| -4 \cos x |$ equals

(A) $\dfrac{4}{3}$ (D) 12

(B) $4 \sin x$ (E) $4 \tan x$

(C) $\dfrac{3}{4}$

29. A right circular cylinder has height h, diameter $\frac{1}{2}h$, and volume 5. What is h?

(A) 6.8   (D) 2.94

(B) 2.5   (E) 4

(C) $\dfrac{3}{4}$

30. One card is drawn from a standard deck of cards. What is the probability of an ace being drawn?

(A) $\dfrac{1}{10}$ (D) $\dfrac{4}{13}$

(B) $\dfrac{12}{13}$ (E) $\dfrac{1}{52}$

(C) $\dfrac{1}{13}$

31. Find the minimum value of the function $f(x) = (x - 1)^2 + 3$.

(A) 0   (D) 3

(B) 1   (E) -2

(C) 2

32. If $f(x) = 2x - 5$ and $g(x) = 4x$, then $f(f(-x) - g(x))$ is given by:

(A) $-12x - 10$ (D) $-6x - 5$

(B) $-12x - 5$ (E) $-4x - 10$

(C) $-12x - 15$

33. If x is an angle with $0 \le x \le 2\pi$, then $-\sec(-x)$ is equal to:

(A) $-\sin x$ (D) $\dfrac{-1}{\cos x}$

(B) $1 - \csc x$ (E) $\dfrac{1}{\cos x}$

(C) $1 - \sec x$

34. For $\triangle ABC (\overline{AB} = 3, \overline{AC} = 4, \overline{BC} = 5)$ the measures of angles a, b, and c are respectively:

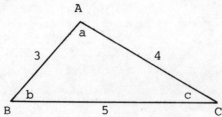

(A) $90°$, $53°$, and $37°$ (D) $100°$, $50°$, and $30°$

(B) $90°$, $60°$, and $30°$ (E) $100°$, $53°$, and $27°$

(C) $75°$, $60°$, and $45°$

35. If x varies inversely as the root of y, and if $x = 5$ and $y = 7$, the constant of variation is

(A) 13.2 (D) 9.6

(B) 23.8 (E) 1.4

(C) 35.0

36. If $f(x) = 3x - 5$ and $g(f(x)) = x$, then $g(x) =$

   (A) $\dfrac{x-5}{3}$

   (D) $\dfrac{x+5}{4}$

   (B) $\dfrac{x+5}{3}$

   (E) $\dfrac{5-x}{3}$

   (C) $\dfrac{2x+5}{3}$

37. If $f(x) = \frac{1}{x}$ and $g(x) = x^2 - 1$ then for which values of x is $f(g(x))$ undefined?

   (A) 0

   (D) all values of x

   (B) 1

   (E) no values of x

   (C) ±1

38. A cube has its length, width, and height all equal to 5. The length of its diagonal is

   (A) 11.2

   (D) 9.1

   (B) 5.22

   (E) 7.2

   (C) 8.66

39. For $0 \le x \le \pi$, the x-intercept of the equation $y = \frac{1}{2} + \cos 2x$ is the point $x =$

   (A) $\dfrac{\pi}{3}$

   (D) 0

   (B) $\dfrac{-4\pi}{3}$

   (E) $\dfrac{-5\pi}{3}$

   (C) $\dfrac{4\pi}{3}$

40. If the functions f, g, and h are defined by

   $f(x) = x^3$, $g(x) = x - 2$, and $h(x) = \dfrac{x}{2}$, then $f(g(h(8))) =$

   (A) 0

   (B) 4

(C)   8

(D)   27

(E)   62

41.   Solve the following inequality:

$(x - 3)^2 (x - 6) > 0.$

(A) $x < 3$ and $3 < x < 6$

(D) $x \geq 6$ and $x = 3$

(B) $x \geq 6$

(E) $x = 6, x = 3$

(C) $x > 6$

42.   Find the coordinates of the turning point of the parabola

$y = x^2 - 8x + 15.$

(A)   (2, 0)

(D)   (5, 0)

(B)   (0, –1)

(E)   (–4, 1)

(C)   (4, –1)

43.   Which of the graphs below represents the equation

$\dfrac{x^2}{9} - \dfrac{y^2}{9} = 1?$

(A)

(B)

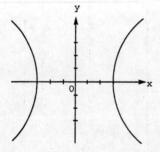

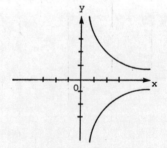

(C)

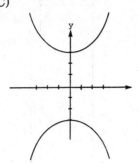

(D)

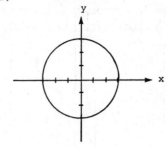

(E)

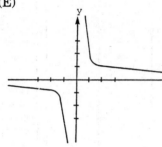

44. What is the radian measure for the angle of 12°?

(A) 0.5678

(D) 0.2501

(B) 1.1209

(E) 0.1504

(C) 0.2094

45. Given $2\tan^2 x + \sec^2 x = 2$, what is $\tan x$?

(A) $\frac{1}{2}\sqrt{3}$

(D) $2\sqrt{3}$

(B) $\frac{1}{\sqrt{3}}$

(E) $\pm\sqrt{3}$

(C) $\pm\frac{1}{\sqrt{3}}$

46. Find the smallest of 3 consecutive positive integers such that when 5 times the largest is subtracted from the square of the middle one, the result exceeds three times the smallest by 7.

(A) 9

(B) 10

(C) 8

(D) 2

(E) 6

47. What is the area of a right triangle with an angle of 35° and with the length of the longer leg 7?

(A) 18

(B) 30

(C) 24

(D) 17

(E) 34

48. Solve for x, when $|\ 5 - 3x\ | = -2$.

(A) $\dfrac{7}{3}$

(B) $-\dfrac{7}{3}$

(C) -1

(D) 1

(E) No solution

49. The graph $f(x) = 4x^2 - 3.5$ passes the x-axis at

(A) 0.5

(B) 3.5

(C) 4.0

(D) 0.935

(E) 0.875

50. Let n be an integer greater than or equal to 1. Which of the following is greater than 1?

(A) $-(n)^3$

(B) $(n)^{-\frac{1}{3}}$

(C) $-(-n)^3$

(D) $-(n + 1)^3$

(E) $(-n + 1)^5$

SAT Mathematics Level 2

Practice Test 4
ANSWER KEY

1.	(A)	14.	(B)	27.	(E)	40.	(C)
2.	(A)	15.	(C)	28.	(A)	41.	(C)
3.	(C)	16.	(B)	29.	(D)	42.	(C)
4.	(B)	17.	(A)	30.	(C)	43.	(A)
5.	(B)	18.	(C)	31.	(D)	44.	(C)
6.	(B)	19.	(C)	32.	(C)	45.	(C)
7.	(C)	20.	(B)	33.	(D)	46.	(C)
8.	(E)	21.	(B)	34.	(A)	47.	(D)
9.	(B)	22.	(A)	35.	(A)	48.	(E)
10.	(B)	23.	(B)	36.	(B)	49.	(D)
11.	(B)	24.	(A)	37.	(C)	50.	(C)
12.	(A)	25.	(A)	38.	(C)		
13.	(A)	26.	(A)	39.	(A)		

DETAILED EXPLANATIONS
OF ANSWERS

1. **(A)**

"Total surface area is 24" implies that the area of one face is

$$\frac{24}{6} = 4 \text{ cm}^2.$$

Hence the length of one side is 2 cm which implies the volume is

$$2^3 = 8 \text{ cm}^3.$$

2. **(A)**

The probability of obtaining "even" or "greater than 3" is expressed below:

$$P(\text{even or} > 3) = P(\text{even}) + P(>3) - P(\text{even} \cap > 3)$$

$$= \frac{3}{6} + \frac{3}{6} - \frac{2}{6} = \frac{4}{6} = \frac{2}{3}$$

Note that $P(\text{even} \cap >3)$ means "the probability of obtaining a number which is even and greater than 3 at the same time." Out of six possibilities only two, 4 and 6, are even and greater than 3.

3. **(C)**

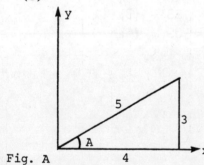

Fig. A

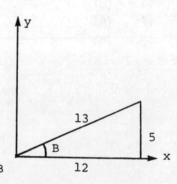

Fig. B

Since $\tan A = {}^3/_4$, then by the Pythagorean Theorem $\sin A = {}^3/_5$ and $\cos A = {}^4/_5$ (see Figure A). Also $\sin B = {}^5/_{13}$, since $\cos B = {}^{12}/_{13}$ (see Figure B).
Now

$$\sin(A + B) = \sin A \cos B + \cos A \sin B$$

$$= \left(\frac{3}{5}\right)\left(\frac{12}{13}\right) + \left(\frac{4}{5}\right)\left(\frac{5}{13}\right)$$

$$= \frac{36}{65} + \frac{20}{65} = \frac{56}{65}.$$

4.   **(B)**
You have to solve this problem in two steps. First, use the log function to change the equation $13^x = 15$ into a solvable form, i.e.,

$$\log(13^x) = x\log(13) = \log(15)$$

So, $x = \dfrac{\log(13)}{\log(15)}$

Calculator: $\log 13 \div \log 15 = 1.05579$. That is $x = 1.05579$.

Then, go on to the second step to calculate x^5 since x is now known.

$$1.05579 \times 1.05579 \times 1.05579 \times 1.05579 \times 1.05579 = 1.311787.$$

5.   **(B)**
The midpoint $M(x, y)$ of the line segment is given by the relations

$$x = \frac{1}{2}(x_1 + x_2), \ y = \frac{1}{2}(y_1 + y_2)$$

Thus, the coordinates of the midpoint are:

$$x = \frac{1}{2}(1 + 5) = 3, \ y = \frac{1}{2}(2 + 8) = 5$$

$\therefore$ Coordinates = (3, 5).

6.   **(B)**
This is an inverse function problem. So, you can put $f(x) = 7$ into the equation and find x. Specifically,

$$7 = 5^x + 4$$

This leads us to use the log function to simplify the problem. That is,

$$\log(7 - 4) = \log(5^x) = x\log(5)$$

Then, we have

$$x = \frac{\log 3}{\log 5} = 0.6826.$$

7. **(C)**
    Rewrite the inequality as

    $$y > \frac{x}{2} + 3.$$

To determine the region of solution, graph the line

$$y = \frac{x}{2} + 3.$$

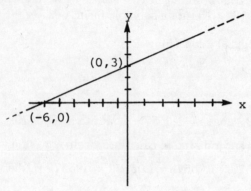

The slope is $\frac{1}{2}$ and the y-intercept is 3. Now the solution of the inequality is the set of all points (x, y) on the $x - y$ plane whose y-coordinates are greater than $x/2 + 3$, that is, above the line $y = x/2 + 3$. We see from the graph that this region lies in quadrants I, II, and III. A fast method to determine the answer when we have a first order double variable inequality is to draw the line whose equation is found by replacing the inequality sign by that of equality, and then to check whether the origin $(0, 0)$ holds in the inequality. If it does, then the solution set of the inequality is the semi-plane whose border is the line and contains the origin. If it does not, the solution is the other semi-plane.

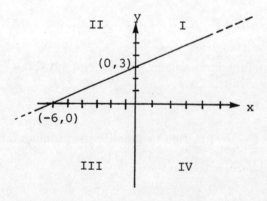

8. **(E)**
    If we open up the figure we can see that the total surface area A is given by:

    $$2A_1 + A_2 + A_3 + A_4 + A_5$$

$$A_1 = 5 \times 4 + \frac{4 \times 3}{2}$$

$$= 20 + 6 = 26$$

$$A_2 = 20$$

$$A_3 = 20$$

$$A_4 = 32$$

$$A_5 = 16$$

$$A = 2 \times 26 + 20 + 20 + 32 + 16 = 140$$

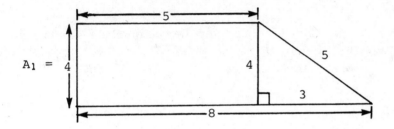

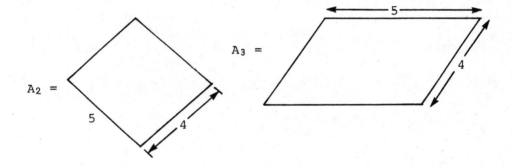

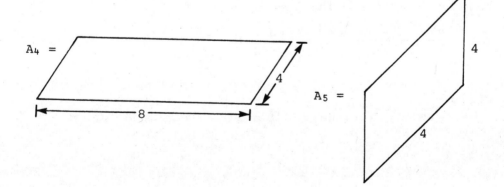

9. **(B)**

$$\left(\frac{27}{35}\right)^{-\frac{2}{3}} = \frac{1}{\left(\frac{27}{35}\right)^{\frac{2}{3}}} = \frac{1}{\sqrt[3]{\left(\frac{27}{35}\right)^2}} = \frac{1}{\sqrt[3]{0.77^2}} = 1.19$$

10. **(B)**

 x can be replaced in the formula

$$\frac{x}{|x|}$$

with any real number except 0, so if x is negative,

$$\frac{x}{|x|} = -1,$$

and if x is positive, $x = 1$. Thus, there are only the two numbers -1 and 1 in the range of our function: $R = \{-1, 1\}$.

11. **(B)**
 Multiplying

$$\frac{n!}{(n+1)!}$$

by

$$\frac{n+2}{n+2}$$

gives

$$\frac{n!(n+2)}{(n+2)!}.$$

Therefore, the problem becomes

$$\frac{n!(n+2)}{(n+2)!} + \frac{(n+1)!}{(n+2)!}$$

where $(n + 1)! = n! (n + 1)$. This gives us

$$\frac{n![(n+2) + (n+1)]}{(n+2)!} = \frac{2n+3}{(n+1)(n+2)}$$

12. **(A)**
 Let $\theta = \arcsin \frac{1}{2}$. Then $\sin\theta = \frac{1}{2}$. Draw a triangle with $\sin\theta = \frac{1}{2}$:

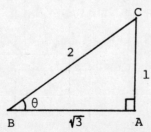

By the Pythagorean Theorem, the side adjacent to θ is $\sqrt{3}$. These are the sides of a 30°–60°–90° triangle. Hence $\theta = 30°$ and

$$\cos(\arcsin\frac{1}{2}) = \cos 30° = \frac{\sqrt{3}}{2}.$$

Also without evaluating the angle, we could use the definition of cosine in the above triangle and write:

$$\cos\theta = \frac{AB}{BC} = \frac{\sqrt{3}}{2}$$

13.   (A)

First, calculate $g(2)$:

$$g(2) = 2^3 + 2 = 8 + 2 = 10$$

Thus,

$$f(g(2)) = f(10) = \sqrt{10+1} = \sqrt{11} = 3.3166$$

14.   (B)

Use synthetic division:

$$
\begin{array}{r}
2x - 6 \\
x+1 \overline{\smash{\big)}\ 2x^2 - 4x + 5} \\
\underline{-(2x^2 + 2x)} \\
-6x + 5 \\
\underline{-(-6x - 6)} \\
11
\end{array}
$$

The remainder is 11.

15.   (C)

Since Sir Lancelot must sit in an assigned chair, and Sir Gawain on either side of him, there are 4! or 24 ways of seating the other 4. For each of these combinations, Sir Gawain can be in either of two seats, so the total number of ways of seating the knights is 24×2 or 48.

16.   (B)

An odd function $f(x)$ has the property that whenever $x \in$ domain of f, then we have $(-x) \in$ domain and also:

$$f(-x) = -f(x).$$

All of the functions below have the real number system as their domain, i.e., they are defined for any real number. Therefore the first condition for being an odd function is satisfied. But for the second condition:

I. $f(-x) = -2x + 1$, $-f(x) = -2x - 1$.

These are not equal.

II. $\tan(-x) = \dfrac{\sin(-x)}{\cos(-x)} = \dfrac{-\sin x}{\cos x} = -\tan x$

Thus, $g(x) = \tan x$ is an odd function.

III. $h(-x) = -x^3 + x^2$, $-h(x) = -x^3 - x^2$.

These are not equal.
     Hence, the only odd function is $g(x) = \tan x$.

17.   **(A)**

$$\frac{n!}{n(n+1)} = \frac{n \times (n-1) \times (n-2)...1}{n(n+1)}$$

$$= \frac{(n-1)(n-2)...1}{(n+1)}$$

$$= \frac{(n-1)!}{(n+1)}$$

18.   **(C)**

$$\tan\theta\,(x + \cot\theta\,\cos\theta) = \sec\theta$$

Rewrite in terms of $\sin\theta$ and $\cos\theta$:

$$\frac{\sin\theta}{\cos\theta}\left(x + \frac{\cos\theta}{\sin\theta}\cos\theta\right) = \frac{1}{\cos\theta}$$

$$(\sin\theta)x + \cos^2\theta = 1$$

$$\Rightarrow \qquad\qquad x = \sin\theta.$$

19.   **(C)**
     Assume $x = \log_3 6$. The question is how to calculate x. Remember, the logarithm of the base itself equals one. By using this fact, we can have

$$x\log_3 3 = \log_3 6$$

which gives us $3^x = 6$. So now the problem becomes finding the solution for equation $3^x = 6$.

Calculator: This can be done very easily, because now you can use the log function on your calculator to solve this problem. That is by taking $\log 3^x = \log 6$, you have

$$x = \frac{\log 6}{\log 3} = 1.6309$$

20. **(B)**

$$f(g(h(\pi))) = f\left(g\left(\frac{\pi}{2}\right)\right)$$

$$= f\left(\sin\frac{\pi}{2}\right)$$

$$= f(1) = 1^2 = 1,$$

since we have

$$h(\pi) = \frac{\pi}{2} \text{ and } g\left(\frac{\pi}{2}\right) = \sin\frac{\pi}{2} = 1.$$

21. **(B)**

Distance d between any two points (x_1, y_1), (x_2, x_2) in the two-dimensional coordinate system is

$$d = \sqrt{(x_1 - x_2)^2 + (y_1 - y_2)^2}$$

The problem can be calculated as:

$$d = \sqrt{(2 - (-3))^2 + (5 - (-3))^2} = \sqrt{89} = 9.434$$

22. **(A)**

$$0.18\overline{18} = 0.18 + 0.0018 + \ldots = \sum_{n=0}^{\infty} 0.18(0.01)^n$$

which is of the form

$$\sum_{n=0}^{\infty} ar^n$$

with $a = 0.18$, $r = 0.01$. This is a geometric series. The sum is

$$\frac{a}{1-r} = \frac{0.18}{1-0.01} = \frac{0.18}{0.99} = \frac{2}{11}.$$

23.   **(B)**

There are:

$\binom{8}{2} = 28$ and

$\binom{5}{2} = 10$ ways

of choosing officers with the criteria stated.

TOTAL $= 28 \times 10 = 280$ total ways of choosing officers.

24.   **(A)**

$$\lim_{x \to 0}\left(\frac{x^2 - 2x}{x}\right) = \lim_{x \to 0}\left(\frac{x^2}{x} - \frac{2x}{x}\right)$$

$$= \lim_{x \to 0}(x - 2)$$

$$= 0 - 2 = -2$$

Note that we could not set $x = 0$ before simplifying the fraction.

25.   **(A)**

As shown in the figure, the line

$$y = 2x + 3$$

passes the x-axis and the y-axis at $(-1.5, 0)$ and $(0, 3)$, respectively. So, the triangle has legs with lengths 1.5 and 3. Thus, we calculate its area by

$$\frac{1}{2} \times 1.5 \times 3 = 2.25$$

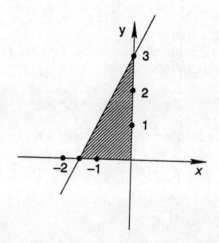

26.   **(A)**

$$f(x) \times g(x) = x^3(x^2 + 1) = x^5 + x^3.$$

This is an odd function since $f(-x) = -f(x)$.

$$f(g(x)) = f(x^2 + 1) = (x^2 + 1)^3.$$

This is not odd because x is squared first, then cubed. Positive and negative values of x yield the same result when raised to an even power.

$$2g(f(x)) - 1 = 2g(x^3) - 1$$

$$= 2(x^3)^2 + 2 - 1$$

$$= 2x^6 + 1.$$

Again, x is raised to an even power so the function is not odd.

27.   **(E)**
Complete the squares:

$$x^2 + 4x - 2y + y^2 + 2 = 0$$

$$(x^2 + 4x + 4) - 4 + (y^2 - 2y + 1) - 1 + 2 = 0$$

$$(x + 2)^2 - 4 + (y - 1)^2 - 1 + 2 = 0$$

$$(x + 2)^2 + (y - 1)^2 = 3 = (\sqrt{3})^2$$

This has the form

$$(x - h)^2 + (y - k)^2 = r^2,$$

where r is the radius. Hence $r = \sqrt{3}$.

Note that any quadratic two-variable equation in which the coefficient of x^2 is equal to that of y^2, represents a circle, where the center can be anywhere on the x–y plane.

28.   **(A)**
Since $\sin x = \dfrac{1}{3}\tan x$, then

$$\sin x = \frac{1}{3}\frac{\sin x}{\cos x}.$$

Dividing throughout by $\sin x$, we have

$$1 = \frac{\frac{1}{3}}{\cos x} \text{ or } \cos x = \frac{1}{3}$$

$$|-4\cos x| = \left|-4\left(\frac{1}{3}\right)\right| = \frac{4}{3}$$

29.   **(D)**
The radius of the cylinder is $^1\!/_4 h$. So, the volume of the cylinder is

$$\left(\frac{1}{4}h\right)^2 \pi \times h$$

which equals 5 as given in the problem. Thus we can solve for the parameter h from the equation:

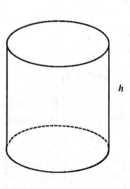

h

$$h = \sqrt[3]{\frac{5 \times 16}{\pi}} = 2.94$$

30.   (C)

There are 4 aces in a deck of 52 cards. Hence the probability of drawing an ace is

$$\frac{4}{52} = \frac{1}{13}.$$

31.   (D)

Since $(x - 1)^2 > 0$ for all x, the minimum value occurs when $(x - 1)^2 = 0$. Hence,

$$f(x) = 0 + 3 = 3$$

is the minimum value. This conclusion can also be made by considering the graph of the function which is a parabola.

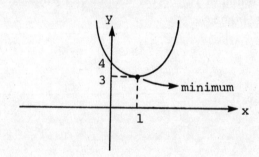

32.   (C)

If $f(x) = 2x - 5$ and $g(x) = 4x$, then

$$f(-x) = 2(-x) - 5 = -2x - 5$$

$$f(-x) - g(x) = -2x - 5 - 4x = -6x - 5$$

$$f(-6x - 5) = 2(-6x - 5) - 5 = -12x - 10 - 5 - 12x - 15$$

33.   (D)

By definition

$$\sec(-x) = \frac{1}{\cos(-x)}.$$

If we multiply both sides by −1, we obtain:

$$-\sec(-x) = \frac{-1}{\cos(-x)}$$

Also $\cos(-x) = \cos x$ (see figure below) so we conclude that

$$-\sec(-x) = \frac{-1}{\cos(-x)} = \frac{-1}{\cos x}$$

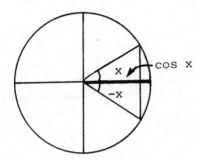

34. **(A)**

By inspection it can be seen that $\triangle ABC$ is a right triangle ($3^2 + 4^2 = 5^2$). $\overline{BC}$ is the hypotenuse and, therefore, angle $a = 90°$.

$$\angle b = \sin^{-1}\left(\frac{4}{5}\right) \approx 53°$$

$$\angle c = \sin^{-1}\left(\frac{3}{5}\right) \approx 37°$$

35. **(A)**

The variations of x and y are constrained by the equation

$$x\sqrt{y} = k,$$

where k is the constant of variation. So, to find k you only need to plug x and y into this equation:

$$x\sqrt{y} = 5\sqrt{7} = 13.22 = k$$

36. **(B)**

$$f(x) = 3x - 5 \text{ and } g(f(x)) = x.$$

We want to know what value of x (in terms of x) would cause $g(3x - 5)$ to equal x. To distinguish one x from the other we can call the x from $3x - 5$ by y. This will result in:

$$3y - 5 = x$$

$$3y = x + 5$$

$$y = \frac{x+5}{3}$$

If we use $g(y) = g(3x - 5) = \dfrac{3x - 5 + 5}{3} = x.$

37.   **(C)**

$$f(g(x)) = f(x^2 - 1) = \frac{1}{x^2 - 1}$$

which is undefined for $x^2 - 1 = 0$ or $x = \pm 1$. Generally, the domain of the function

$$(f \circ g) = f(g(x))$$

consists of those values of the domain of g for which $g(x)$ belongs to the domain of the function f. Therefore we have:

$$\text{Dom}(f \circ g) = \{x \in \text{Dom of } g \mid g(x) \in \text{Dom of } f\}$$

Thus: $\text{Dom}(f \circ g) \subset \text{Dom } g$

38.   **(C)**
    As shown in the figure, the diagonal on each side of the cube is

$$\sqrt{5^2 + 5^2} = 7.07$$

The diagonal of the cube lies in the rectangle whose height is 5 and width is 7.07, as shown in the figure. Therefore, it can be easily found that the length of the diagonal is

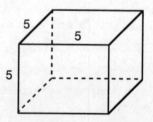

$$\sqrt{5^2 + 7.07^2} = 8.66$$

39.   **(A)**

Let $\frac{1}{2} + \cos 2x = 0$

$\cos 2x = -\frac{1}{2}$

Since $0 \le x \le \pi$ then $0 \le 2x \le 2\pi$

The two angles where the cosine is equal to $-\frac{1}{2}$ on the interval $[0, 2\pi]$ are $\frac{2\pi}{3}$ and $\frac{4\pi}{3}$. Thus,

$2x = \frac{2\pi}{3}$ or $2x = \frac{4\pi}{3}$

$x = \frac{\pi}{3}$ $x = \frac{2\pi}{3}$

In this case, the correct answer choice is $\frac{\pi}{3}$.

40.   **(C)**

$f(g(h(8))) = f(g(4))$

$= f(2) = 2^3 = 8.$

41.   **(C)**

$(x - 3)^2$ is positive for all values of x, hence $(x - 6)$ must also be positive for their product to be positive.

Thus $(x - 3)^2 > 0$ and $x - 6 > 0 \Rightarrow x > 3$ and $x > 6$. The intersection of these two sets is $x > 6$.

42.   **(C)**

$y = x^2 - 8x + 15$

The above equation is of the general quadratic form:

$f(x) = ax^2 + bx + c,$

here $a = 1$, and $b = -8$.

$x = \dfrac{-b}{2a}$

is the equation for the axis of symmetry of the parabola. Hence $x = -8/2 = 4$ is the x value at the turning point.

When $x = 4$, we find y to be:

$$y = 4^2 - 8(4) + 15 = -1.$$

Hence the coordinates of the turning point are $(4, -1)$.

43. **(A)**

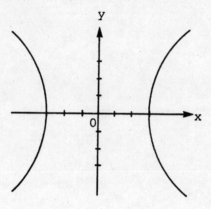

$$\frac{x^2}{9} - \frac{y^2}{9} = 1$$

is an equation of the form:

$$\frac{x^2}{a^2} - \frac{y^2}{b^2} = 1$$

with $a = 3$ and $b = 3$.

Therefore the graph is a hyperbola. The x-intercepts are found by setting $y = 0$:

$$\frac{x^2}{9} - \frac{0^2}{9} = 1$$

$$x^2 = 9$$

$$x = \pm 3.$$

Thus, the x-intercepts are at $(-3, 0)$ and $(3, 0)$. There are no y-intercepts since for $x = 0$ there are no real values of y satisfying the equation, i.e., no real value of y satisfies

$$\frac{2}{-1+1} - 3 = \frac{4(-1)+6}{-1+1}$$

$$\frac{2}{0} - 3 = \frac{-4+6}{0}$$

Since division by zero is impossible the above equation is not defined for $x = -1$. Hence we conclude that the equation has no roots for $x = -1$.

$$\left\{ x \left| \frac{2}{x+1} - 3 = \frac{4x+6}{x+1} \right. \right\} = \phi$$

44.   (C)
    180° corresponds to π in the radian measure. So, for each degree of angle, we can have $\pi/_{180}$ amount of angle in the radian measure. Thus, 12° should have radian measure

$$12 \times \frac{\pi}{180} = 0.2094$$

Calculator: First, set your calculator to the degree mode and do sin(12). You should get 0.2079. Then change your calculator to the radian mode and do the inverse sin on 0.2079, that is sin⁻¹(0.2079). Since the calculator is now in the radian mode, it will give you the answer 0.2094.

45.   (C)
    We are given

$$2\tan^2 x + \sec^2 x = 2.$$

Using the fact that $\sec^2 x = 1 + \tan^2 x$ we have

$$2\tan^2 x + (1 + \tan^2 x) = 2 \text{ or } 3\tan^2 x = 1$$

$$\Rightarrow \qquad \tan x = \frac{\pm 1}{\sqrt{3}}.$$

46.   (C)
    Let x be the smallest number; this implies that the three consecutive numbers are x, $x + 1$, and $x + 2$. Therefore

$$(x + 1)^2 - 5(x + 2) = 3x + 7$$

$$x^2 + 2x + 1 - 5x - 10 = 3x + 7$$

which leads to

$$x^2 - 6x - 16 = 0$$

or $$(x - 8)(x + 2) = 0$$

$$\Rightarrow \qquad x = -2 \text{ or } 8.$$

But x cannot be a negative integer, so $x = 8$ is the answer.

47.   **(D)**
First, find the length of the other leg of the triangle. Because $35° < 45°$, the leg next to the given angle should be the longer one. So, the length of the leg is:

$$7\tan(35°) = 4.9$$

Finally, since the area of a right triangle is

$$\frac{l_1 \times l_2}{2},$$

we have

$$\frac{4.9 \times 7}{2} = 17.155$$

Round this to 17.

48.   **(E)**
This problem has no solution, since the absolute value can never be negative.

49.   **(D)**
The roots of the equation $4x^2 - 3.5$ are points on the x-y plane where the graph passes the x-axis. The roots of the equation can be found by

$$x = \sqrt{\frac{3.5}{4.0}} = \pm 0.935$$

50.   **(C)**
A negative integer raised to any odd power is negative; the negative sign before the parentheses in $-(-n)^3$ makes the negative value of $(-n)^3$ always positive.

THE SAT SUBJECT TEST IN

Math
Level 2

PRACTICE TEST 5

SAT Mathematics
Level 2

Practice Test 5

Time: 1 Hour
50 Questions

DIRECTIONS: Choose the best answer for each question and mark the letter of your selection on the corresponding answer sheet in the back of the book.

NOTES:

(1) Some questions require the use of a calculator. You must decide when the use of your calculator will be helpful.

(2) You may need to decide which mode your calculator should be in—radian or degree.

(3) All figures are drawn to scale and lie in a plane unless otherwise stated.

(4) The domain of any function f is the set of all real numbers x for which $f(x)$ is a real number, unless other information is provided.

REFERENCE INFORMATION: The following information may be helpful in answering some of the questions.

Volume of a right circular cone with radius r and height h	$V = \dfrac{1}{3}\pi r^2 h$
Lateral area of a right circular cone with circumference of the c and slant height l	$S = \dfrac{1}{2}cl$
Volume of a sphere with radius r	$V = \dfrac{4}{3}\pi r^3$
Surface Area of a sphere with r	$S = 4\pi r^2$
Volume of a pyramid with base area B and height h	$V = \dfrac{1}{3}Bh$

1. Determine the surface area of the figure below:

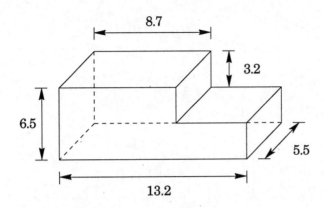

(A)   309.55   (D)   453.55

(B)   363.35   (E)   471.15

(C)   359.50

2. The following two figures have the same area. What is the value of x?

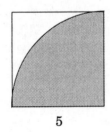 

5   x

(A)   5.21   (D)   8.86

(B)   2.80   (E)   4.43

(C)   6.51

3. Given $f(x) = \dfrac{2x + 4}{2}$, determine $g(x)$ if $g(f(x)) = 2x$.

(A) $g(x) = 2x - 2$ (D) $g(x) = 3x$

(B) $g(x) = x - 2$ (E) $g(x) = 4x + 4$

(C) $g(x) = 2x - 4$

4. Given $f(x) = \sqrt{2^x - 1}$ and $g(x) = x^2 - 1$, determine $f(-g(-x))$.

   (A) $\sqrt{2^{x^2-1}-1}$

   (D) $\sqrt{2^{x^2-1}}$

   (B) $\sqrt{2^{-x^2+1}-1}$

   (E) $\sqrt{2^{-x^2+1}}$

   (C) $\sqrt{2^{-x^2+1}+1}$

5. For the triangle below, determine $\sin 2\alpha$.

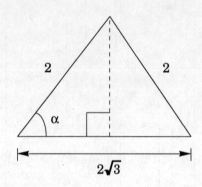

   (A) 0.5

   (D) -0.5

   (B) 0

   (E) 0.866

   (C) 0.577

6. If $3x^2 - 4x = 55$, what is the positive value of x?

   (A) 4

   (D) 5.5

   (B) 5

   (E) 6

   (C) 3

7. If $4x^{\frac{3}{2}} + 5x^{\frac{2}{3}} = 19.25$, then $x =$

   (A) 5

   (D) 7

   (B) 3

   (E) 4

   (C) 2

8. Given $z = -3 + 2i$, determine z^5. Assume $i = \sqrt{-1}$ and $i^2 = -1$

(A) $597.15 + 121.87i$ (D) $-121.87 + 597.15i$

(B) $-597.15 - 121.87i$ (E) $121.87 - 597.15i$

(C) $-597.15 + 121.87i$

9. If $f(x)$ is $3x^4 - 2x^3 + x^2$ and $g(x)$ is $x^5 - 2x^2$, then the order of the product $f(x) \times g(x)$ is

(A) 4   (D) 2

(B) 5   (E) 9

(C) 3

10. Find the radius of the circle which has the following equation

$$x^2 + y^2 + 6x + 8y - 11 = 0$$

(A) 6   (D) 12

(B) 8   (E) $6\sqrt{2}$

(C) $8\sqrt{2}$

11. Which one of these graphs could represent the system of equations below?

$$\begin{cases} x^2 + y^2 = 49 \\ x^2 + y^2 - 6x - 8y + 21 = 0 \end{cases}$$

(A)

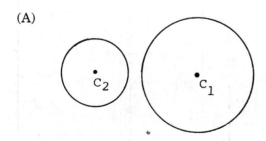

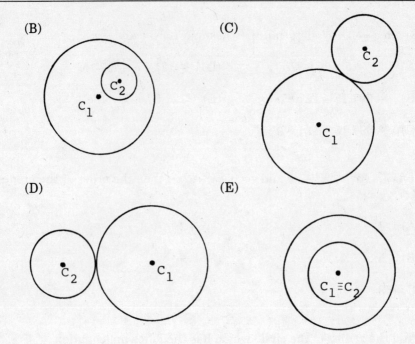

(B)

(C)

(D)

(E)

12. Which one of the graphs below represents the inverse of the function

$y = 3x + 4$?

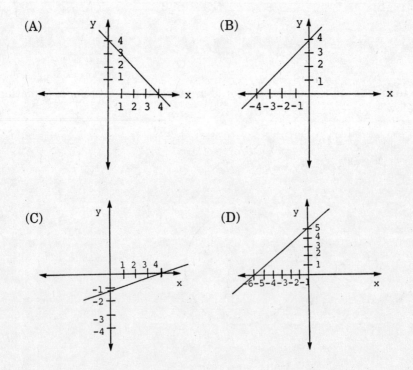

(A)

(B)

(C)

(D)

(E)

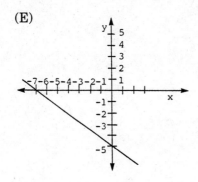

13. Suppose x is a prime number less than 12. What is the probability that $2 \le x \le 6$?

(A)   2:5

(D)   3:4

(B)   1:2

(E)   1:3

(C)   3:5

14. For what value of c will the graph of the function $y = 2x^2 + 5x + c$ NOT intersect the x-axis?

(A)   4

(D)   1

(B)   −1

(E)   2

(C)   0

15. If we toss a coin 5 times, the probability of obtaining exactly 1 head is:

(A) $\dfrac{1}{32}$

(D) $\dfrac{4}{32}$

(B) $\dfrac{2}{32}$

(E) $\dfrac{5}{32}$

(C) $\dfrac{3}{32}$

16. If $\tan^2(3x) = 3$ and $0 \le x \le \dfrac{\pi}{4}$, then $x =$

   (A) $\dfrac{\pi}{24}$ (D) $\dfrac{\pi}{9}$

   (B) $\dfrac{\pi}{18}$ (E) $\dfrac{\pi}{12}$

   (C) $\dfrac{\pi}{20}$

17. Which of the following is a graph of the solution of the inequality $-x^2 + 5x - 6 < 0$?

   (A)   (D)

   (B)   (E)

   (C)

18. If

$$f(x) = \frac{1}{\sqrt{x}}, \quad g(x) = \sqrt[3]{\frac{1}{x^2} + \frac{3}{x}} \quad (x \ne 0),$$

then find $f(g(3))$.

   (A)   0.99   (D)   1.04

   (B)   0.98   (E)   0.97

   (C)   1.02

19. Determine which graph corresponds to $y = 3 + \sin 2x$.

(A)

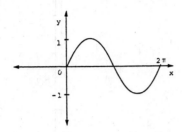

(B)

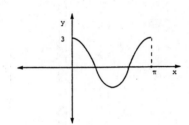

(C)

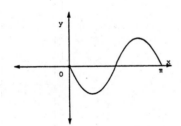

(D)

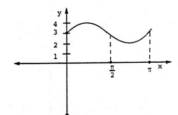

(E)

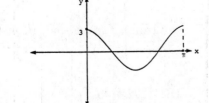

20. Simplify the expression below:

$$\frac{2n!}{(n-1)!} + \frac{2(n-1)!(n-1)!}{(n-2)!^2} =$$

(A) $2(n^2 - n + 1)$

(D) $\frac{2(2n^3 - 7n^2 + 7n - 1)}{n^3 - 5n^2 + 8n - 4}$

(B) $\frac{2n}{n-2}$

(E) $2n + \frac{2}{(n-2)^2}$

(C) $\frac{2n}{(n-1)(n-2)}$

21. Let $f(x) - \ln x + 6$. Determine $\sqrt{\sin(f^{-1}(6))}$.

(A) 0.92

(D) 0

(B) 1

(E) 2.8

(C) -0.92

22. The solution set for

$$\frac{x^2 - 5x + 6}{x - 2} > 0 \text{ is:}$$

(A) $\{x \mid x < 3\}$

(B) $\{x \mid 2 < x < 3 \text{ or } x < -2\}$

(C) $\{x \mid -2 < x < 3\}$

(D) $\{x \mid 2 < x < 3\}$

(E) $\{x \mid x > 3\}$

23. If $\sin x = \frac{1}{2}$ and $\sin y = \frac{\sqrt{2}}{2}$, determine $\tan (x + y)$. (Consider only positive values).

(A) $2 + \sqrt{3}$

(B) $\sqrt{2} + \sqrt{3}$

(C) $2 - \sqrt{3}$

(D) $\dfrac{6 - \sqrt{12}}{4}$

(E) $\dfrac{1}{2}$

24.   Solve for x when $| \, 2x - 2 \, | < 3$.

(A) $\{x \mid x > \frac{5}{2} \text{ or } x < -\frac{1}{2}\}$

(D) $\{x \mid x < -\frac{1}{2}\}$

(B) $\{x \mid x > -\frac{1}{2}\}$

(E) $\{x \mid -2 < x < \frac{5}{2}\}$

(C) $\{x \mid \frac{1}{2} < x < \frac{3}{2}\}$

25.   Determine the value of x in the expression below:

$$\pm\sqrt{\log_{10} 100^x} = 2x$$

(A) $\dfrac{1}{\sqrt{2}}$

(D) 1

(B) 2

(E) $\dfrac{1}{2}$

(C) $\sqrt{2}$

26.   If $f(x) = 1 + \dfrac{1}{2x^3}$, then as x gets very large, $f(x)$ approaches what value in the limit?

(A) 0

(D) 1

(B) $\dfrac{3}{2}$

(E) ∞

(C) $\dfrac{1}{2}$

27.   Determine the area of the triangle with vertices $(4,0)$, $(1, 4)$, and $(5, 1)$.

(A) $\dfrac{7}{2}$

(D) 49

(B) 7

(E) $7\sqrt{2}$

(C) $\dfrac{2}{7}$

28. In triangle ABC with $a = 6$, $b = 5$, $c = 4$, determine the measure of angle A.

   (A) 97.18°   (D) 1.7°

   (B) 1.45°   (E) 47.55°

   (C) 82.8°

29. Determine the quadratic equation whose one root is $12 - 13i$, where $i = \sqrt{-1}$.

   (A) $x^2 - 24x - 313 = 0$ (D) $x^2 - 24x + 313 = 0$

   (B) $x^2 + 24x - 25 = 0$ (E) $x^2 + 26x + 313 = 0$

   (C) $x^2 + 24x + 25 = 0$

30. The expression $(\sin x)\sec x + \cot x$ is equivalent to

   (A) $\cos 2x$ (D) $\tan x + 1$

   (B) $\sec x + 1$ (E) $\dfrac{\cot x + \sec x}{1 - \sin 2x}$

   (C) $\dfrac{2}{\sin 2x}$

31. Define $A * B \overset{\text{def}}{=} A^B$, where A and B are two non-zero integers. Find the value of A if $A = (A * B) * (B * A)$.

   (A) 2   (D) -1

   (B) -2 (E) No possible solution.

   (C) 1

32. Determine the value of the following expression

   $$\log_{10} 10^{3.5} + \ln \frac{1}{5^{-2}} - \log_{10} \sqrt{25} - \ln\left(\frac{27}{30}\right)$$

   (A) -6.6 (D) 2.5

   (B) 5.6   (E) 6.124

   (C) 6.5

33. Find x if the determinant of the matrix A is $2x - 4$, where

$$A = \begin{vmatrix} x-2 & 0 \\ 0 & x-3 \end{vmatrix}$$

(A) $x = 2$ or $x = 5$ (D) $x = 2$ or $x = 3$

(B) $x = 1$ or $x = -1$ (E) $x = -2$ or $x = 1$

(C) $x = 0$ or $x = 2$

34. Determine the coordinates of the foci of the conic with the equation
$(x - 2)^2 + (y - 2)^2 = 1$.

(A) $F_1 = (\sqrt{2}, 0), F_2 = (-\sqrt{2}, 0)$

(B) $F_1 = (0, 1), F_2 = (1, 0)$

(C) $F_1 = (-2, 0), F_2 = (2, 0)$

(D) $F_1 = (2 - \sqrt{2}, 2), F_2 = (2 + \sqrt{2}, 2)$

(E) $F_1 = (-\sqrt{2}, \sqrt{2}), F_2 = (\sqrt{2}, \sqrt{2})$

35. Solve the equation
$\sin^2 3x = 0.75$ for $0 \le x \le 360°$.

(A) $60°, 120°, 240°, 300°$ (D) $20°, 40°, 80°, 100°$

(B) $30°, 150°, 210°, 330°$ (E) $15°, 45°, 75°, 105°$

(C) $10°, 50°, 70°, 110°$

36. Determine q in the equation $4x^2 + qx + 12$ so that the roots r_1 and r_2 satisfy $\dfrac{r_1}{r_2} = 3$.

(A) $0, 16, -16$ (D) -16

(B) $16, -16$ (E) 0

(C) 16

37. Determine the second term of $(-4x - 3y)^4$.

(A) $256x^4$ (D) $1728x^3y$

(B) $768x^3y$ (E) $-1728x^3y$

(C) $256x^2y^2$

38. Suppose the following two statements are true:

"If Jack can run, then Jack can fly."

"Jack can't fly."

Which of the following must be true?

I. "If Jack can fly, then Jack can run."

II. "Jack can't run."

III. "Jack can run."

(A) I only.   (D) I and II only.

(B) II only.   (E) None of the above.

(C) III only.

39. A committee of 5 people is to be selected from a group of 6 men and 9 women. If the selection is made randomly, what is the probability that the committee consists of 3 men and 2 women?

(A) $\dfrac{1}{3}$ (D) $\dfrac{1260}{3003}$

(B) $\dfrac{240}{1001}$ (E) $\dfrac{13}{18}$

(C) $\dfrac{1}{9}$

40. If $f(x) = \sin\,[2(\arcsin x)]$, find $f\!\left(\dfrac{1}{3}\right)$.

(A) $\dfrac{4\sqrt{2}}{9}$ (B) $\dfrac{-4\sqrt{2}}{9}$

(C) $\dfrac{\pm 4\sqrt{2}}{9}$ 　　　　　　(D) $\dfrac{2\sqrt{2}}{9}$

(E) $\dfrac{\pm 2\sqrt{2}}{9}$

41. Solve for x in the following equation:

$x = \sqrt{4x+7}$

(A) $5.32, -1.32$ 　　　　　　(D) $-5.32, -1.32$

(B) $-5.32, 1.32$ 　　　　　　(E) $2+i, 2-i$

(C) $5.32, 1.32$

42. Sum the arithmetic progression

$2 + 4 + 6 + 8 + \dots + 100.$

(A) 5000 　　　　　　(D) 2550

(B) 6000 　　　　　　(E) 4050

(C) 4000

43. If $f(x) = \dfrac{x}{x+1}$, express $f(2x)$ in terms of $f(x)$:

(A) $\dfrac{f(x)}{1-f(x)}$ 　　　　　　(D) $\dfrac{f(x)}{1+2f(x)}$

(B) $\dfrac{2f(x)}{1-f(x)}$ 　　　　　　(E) $\dfrac{2f(x)}{f(x)+1}$

(C) $\dfrac{2f(x)}{2f(x)+1}$

44. Let T be a transformation function, $T(x, y) = (3.5x, 6.3y)$. Suppose the vertices of a rectangle of area 5 undergo the transformation $T(x, y)$. What is the area of the transformed rectangle?

(A) 49 　　　　　　(D) 259.7

(B) 4 　　　　　　(E) 14.8

(C) 110.25

45. Express the number $0.32\overline{32}$ in the form $\dfrac{p}{q}$, where p and q are integers.

(A) $\dfrac{320}{999}$ (D) $\dfrac{32}{99}$

(B) $\dfrac{321}{900}$ (E) $\dfrac{323}{990}$

(C) $\dfrac{323}{900}$

46. If $f(x) = \ln(x^2)$, what is the y-intercept of the graph of $f^{-1}(x)$?

(A) $(0, 0)$ (D) $(0, 1)$

(B) $(0, e)$ (E) $(0, \dfrac{1}{e})$

(C) $(0, \sqrt{e}\,)$

47. If $(\sin x)(\cos x) > 0$ then which of the following must be true?

I. $\tan x > 0$

II. $0 < x < \dfrac{\pi}{2}$ or $\pi < x < \dfrac{3\pi}{2}$

III. $\sec x(\csc x) > 0$

(A) I only.   (D) II and III.

(B) I and II.   (E) I, II, and III.

(C) II only.

48.   Determine the volume of the solid figure below.

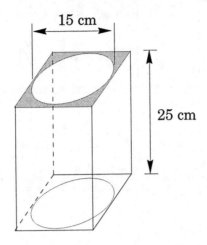

(A)   12046 cm³   (D)   4418 cm³

(B)   4447 cm³   (E)   1207 cm³

(C)   5625 cm³

49.   A certain arc of a circle has a measure of π radians. The ratio between the measure of the arc and the diameter is approximately

(A)   3.14   (D)   1.57

(B)   6.28   (E)   Cannot be determined.

(C)   2

50.   Let $x = t + \dfrac{1}{t}, y = t - \dfrac{1}{t}$. Then the rectangular-coordinate graph of these equations is in the shape of

(A)   an ellipse.   (D)   a circle.

(B)   a straight line.   (E)   a parabola.

(C)   a hyperbola.

SAT Mathematics
Level 2

Practice Test 5
ANSWER KEY

1.	(C)	14.	(A)	27.	(A)	40.	(C)
2.	(E)	15.	(E)	28.	(C)	41.	(A)
3.	(C)	16.	(D)	29.	(D)	42.	(D)
4.	(B)	17.	(D)	30.	(C)	43.	(E)
5.	(E)	18.	(B)	31.	(D)	44.	(C)
6.	(B)	19.	(D)	32.	(E)	45.	(D)
7.	(C)	20.	(A)	33.	(A)	46.	(D)
8.	(C)	21.	(A)	34.	(D)	47.	(E)
9.	(E)	22.	(E)	35.	(D)	48.	(E)
10.	(A)	23.	(A)	36.	(B)	49.	(D)
11.	(B)	24.	(E)	37.	(B)	50.	(C)
12.	(C)	25.	(E)	38.	(B)		
13.	(C)	26.	(D)	39.	(B)		

DETAILED EXPLANATIONS
OF ANSWERS

1. **(C)**

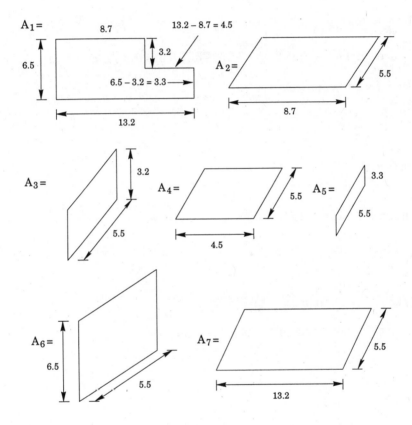

The total surface area is $A = 2A_1 + A_2 + A_3 + A_4 + A_5 + A_6 + A_7$. With the help of a calculator we calculate these values as follows.

Calculator: Before calculating each area press AC to clear the previous entries.

$$A_1 = 6.5 \times 8.7 + (4.5 \times 3.3) = 71.4$$

$$A_2 = 8.7 \times 5.5 = 46.85$$

$$A_3 = 3.2 \times 5.5 = 17.6$$

$$A_4 = 5.5 \times 4.5 = 24.75$$

$$A_5 = 3.3 \times 5.5 = 18.15$$

$$A_6 = 6.5 \times 5.5 = 35.75$$

$$A_7 = 13.2 \times 5.5 = 72.6$$

$$A = (2 \times 71.4) + 46.8 + 17.6 + 24.75 + 18.15 + 35.75 + 72.6$$

$$= 359.50$$

2.   **(E)**

The area of the square is equal to x^2. The area of the quarter disc is $\frac{1}{4} \times \pi \times 5^2$.
If both areas are equal, then

$$x^2 = \frac{1}{4}\pi 5^2$$

$$x = \sqrt{\frac{1}{4}\pi 5^2}$$

Calculator: $1 \div 4 \times \pi \times 5 \text{INV} x^2 = \sqrt{} = 4.4311 \approx 4.43$

3.   **(C)**

If $f(x) = \dfrac{2x+4}{2}$ then $g\!\left(\dfrac{2x+4}{2}\right) = 2x$.
We are looking for a function that transforms

$$\frac{2x+4}{2}$$

into $2x$. If we multiply

$$\left(\frac{2x+4}{2}\right)$$

by 2 and subtract 4 we will have $2x$. This means that the transforming function
is $2x - 4$; so $g(x) = 2x - 4$.

4.   **(B)**

If $g(x) = x^2 - 1$ then

$$g(-x) = x^2 - 1 \text{ and } -g(-x) = -x^2 + 1.$$

If $-g(-x) = -x^2$ then

$$f(-g(-x)) = \sqrt{2^{-x^2+1}} - 1.$$

5.   **(E)**

Since the two sides have the same measure, this is an isosceles triangle and therefore, the altitude bisects the base.

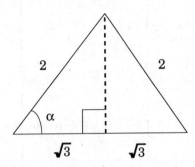

So we conclude that $\cos \alpha = \dfrac{\sqrt{3}}{2}$, or $\alpha = 30°$. Thus

$$\sin 2\alpha = \sin 60° = \dfrac{\sqrt{3}}{2} = 0.866.$$

If the value of α is not known for $\cos \alpha = \dfrac{\sqrt{3}}{2}$, it can be found by calculator as follows:

Calculator: $\sqrt{3} \div 2\, \text{INV COS} = 30°$

$$60 \text{ SIN} = 0.866$$

Be sure that you are working in degrees and not radians.

6.   **(B)**

From the quadratic formula for the equation $ax^2 + bx + c = 0$

$$x = \dfrac{-b \pm \sqrt{b^2 - 4ac}}{2a}$$

In our equation $a = 3$, $b = -4$, $c = -55$, thus

$$x = \dfrac{4 \pm \sqrt{-4^2 - (4)(3)(-55)}}{(2)(3)} = \dfrac{4 \pm \sqrt{16 + 660}}{6} = \dfrac{4 \pm \sqrt{676}}{6}$$

$$= \dfrac{4 \pm 26}{6} = -\dfrac{22}{6}, 5$$

Thus the positive value of x is 5. The value of x can be found by calculator as:

Calculator: $4\, \text{INV} x^2 - (4 \times 3 \times 55 \pm) = +4 = \div 6 = 5$

7.   **(C)**
   Plug in all the possible answers on the left-hand side and check which one gives the correct right-hand side value.

Calculator: 4×2 INV x^y $(\frac{3}{2})$ + $(5 \times 2$ INV x^y $(2 \div 3)) = 19.25$.

Therefore (C) is correct.

8.   **(C)**
   From De Moivre's formula, for a given complex number $z = x + yi$, it is possible to show that

$$z^n = \rho^n (\cos n\,\theta + i\,\sin n\theta)$$

where
$$\rho = \sqrt{x^2 + y^2} = |z| \quad \text{and} \quad \theta = \arctan\left(\frac{y}{x}\right)$$

$$|z| = \sqrt{3^2 + 2^2} = \sqrt{9 + 4} = \sqrt{13}$$

$$\theta = \arctan\left(\frac{2}{3}\right) = 33.69°$$

Calculator: $2 \div 3$ INV TAN = $33.69°$

so
$$z^5 = (\sqrt{13})^5 \,[\cos(5(33.69°)) + i\,\sin(5(33.69°))]$$

Calculator: $33.69 \times \cos = -0.98$

$33.69 \times 5\,\sin = 0.2$

$13\sqrt{}$ INV$xy5 = 609.34$

$609.34 \times (-.98) = -597.15$

$609.34 \times .2 = 121.87$

9.   **(E)**
   Since the order of the product corresponds to the sum of the orders of the polynomials $f(x)$ and $g(x)$, and the order of $f(x)$ is 4 and that of $g(x)$ is 5, the order of their product is 9.

10.   **(A)**
   The circle has the following general equation:
$$x^2 + y^2 + cx + dy + e = 0$$
The radius is given by

$$\gamma = \frac{\sqrt{c^2 + d^2 - 4e}}{2}$$

In our equation $c = 6$, $d = 8$, and $e = -11$. Thus

$$\gamma = \frac{\sqrt{6^2 + 8^2 - 4(-11)}}{2} = \frac{\sqrt{36 + 64 + 44}}{2} = \frac{\sqrt{144}}{2} = \frac{12}{2} = 6$$

Calculator: $6 \text{ INV}x^2 + 8 \text{ INV}x^2 = -(4 \times 11\pm) = \sqrt{} \div 2 = 6$

11.   **(B)**
   The first equation $x^2 + y^2 = 49$ corresponds to a circle with center $C_1 = (0, 0)$ and radius 7.
   The second equation $x^2 + y^2 - 6x - 8y + 21 = 0$ can be rewritten as follows:

$$x^2 - 6x + 9 - 9 + y^2 - 8y + 16 - 16 + 21 = 0$$

The equation above can be rearranged:

$$(x - 3)^2 + (y - 4)^2 = 4 = 2^2$$

   This equation represents a circle with center $C_2 = (3, 4)$ and radius 2.
   Since C_2 is inside the first circle, alternatives (A), (C), and (D) can be eliminated.
   The centers C_1 and C_2 do not coincide, eliminating alternative (E). The only remaining alternative is (B).

12.   **(C)**
   The inverse of the function $f(x) = 3x + 4$ is obtained by replacing y by x and x by y.

$$y = 3x + 4$$

$$x = 3y + 4 \Rightarrow 3y = x - 4 \Rightarrow y = \frac{x - 4}{3}$$

The graph of $\dfrac{x - 4}{3}$ is given below:

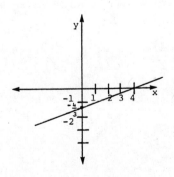

13.   **(C)**

There are 5 prime numbers less than 12:

2, 3, 5, 7, 11.

there are 3 which fall between 2 and 6, namely 2, 3, and 5. So the probability is $\dfrac{3}{5}$.

14.   **(A)**

The function $y = 2x^2 + 5x + c$ can be represented graphically as a parabola.

If its discriminant $D = b^2 - 4ac$ is negative, then the parabola will not intersect the x-axis. This occurs because the equation $0 = 2x^2 + 5x + c$ will have no real solutions when $D < 0$. Setting $D < 0$, we obtain:

$$D = b^2 - 4ac = 25 - 4(2)\,c < 0$$

$$25 - 8c < 0$$

$$25 < 8c$$

$$\frac{25}{8} < c$$

$$c > \frac{25}{8}$$

so the only choice among those given is $c = 4$.

15.   **(E)**

The probability of obtaining a head the first time and tails the other four times is:

$$\underset{p(\text{head})}{\left(\frac{1}{2}\right)} \times \underset{p(\text{tail})}{\left(\frac{1}{2}\right)} \times \underset{p(\text{tail})}{\left(\frac{1}{2}\right)} \times \underset{p(\text{tail})}{\left(\frac{1}{2}\right)} \times \underset{p(tail)}{\left(\frac{1}{2}\right)}$$

Since we could also have obtained a head on the second, third, fourth, or fifth time we played (that is, the head can come up in any of 5 tosses), the probability is

$$5 \times \left(\frac{1}{2}\right) \times \left(\frac{1}{2}\right) \times \left(\frac{1}{2}\right) \times \left(\frac{1}{2}\right) \times \left(\frac{1}{2}\right) = \frac{5}{32}$$

16.   **(D)**

$$\tan^2 (3x) = 3$$

$$\tan (3x) = \pm \sqrt{3}$$

$$3x = \arctan (\pm \sqrt{3})$$

$$x = \frac{\arctan(\pm\sqrt{3})}{3}$$

Calculator: $3 \sqrt{}$ INV TAN $\div 3 = 20°$

or $3 \sqrt{} \pm$ INV TAN $\div 3 = -20°$

Since π rad $= 180°$, therefore

$$20° = \pi \frac{20}{180} = \frac{\pi}{9} \text{ rad}$$

and $-20° = -\frac{\pi}{9}$ rad

since $0 \le x \le \frac{\pi}{4}$, we have $x = \frac{\pi}{9}$.

17.   **(D)**

In order to solve the inequality we can first solve the equation

$$-x^2 + 5x - 6 = 0,$$

as shown below:

$$x = \frac{-(5) \pm \sqrt{(5)^2 - 4(-1)(-6)}}{-2},$$

so $x = \frac{-5-1}{-2} = 3$ or $\frac{-5+1}{-2} = 2$

The graph of the function is shown on the following page.

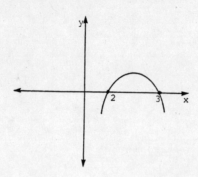

Since we need the values of x that will make the function become negative, the solution is represented below:

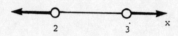

18. **(B)**

$$g(3) = \sqrt[3]{\frac{1}{3^2} + \frac{3}{3}} = \sqrt[3]{\frac{1}{9} + 1} = \sqrt[3]{\frac{10}{9}}$$

$$f(g(3)) = \frac{1}{\sqrt{\sqrt[3]{\frac{10}{9}}}} = \frac{1}{\left(\frac{10}{9}\right)^{\frac{1}{3} \times \frac{1}{2}}} = \frac{1}{\left(\frac{10}{9}\right)^{\frac{1}{6}}}$$

Calculator: $10 \div 9 = \text{INV } x^y \ (1 \div 6) = \text{INV } 1 \div x = 0.98.$

19. **(D)**

Since the function is $3 + \sin 2x$, we find the period first, which is $\dfrac{2\pi}{2} = \pi$. After drawing a function $\sin x$ with period π we slide it upwards by 3 units.

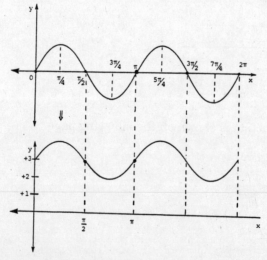

20. **(A)**

$$\frac{2n!}{(n-1)!} + \frac{2(n-1)!(n-1)!}{(n-2)!^2}$$

$$= \frac{2(n)(n-1)!}{(n-1)!} + \frac{2(n-1)(n-2)!(n-1)(n-2)!}{(n-2)!(n-2)!}$$

$$= 2n + 2(n-1)^2$$

$$= 2n + 2(n^2 - 2n + 1)$$

$$= 2n + 2n^2 - 4n + 2$$

$$= 2n^2 - 2n + 2$$

$$= 2(n^2 - n + 1)$$

21. **(A)**

Replacing $f(x)$ by x and x by $f^{-1}(x)$, we get

$$x = \ln f^{-1}(x) + 6 \quad \text{or} \quad \ln f^{-1}(x) = x - 6$$

or $\qquad f^{-1}(x) = ex - 6$

$$\sqrt{\sin(f^{-1}(6))} = \sqrt{\sin(e^0)} = \sqrt{\sin(1)} = 0.917$$

Calculator: $\quad 1 \text{ SIN } \sqrt{} = 0.917 \approx 0.92$

Be sure that you are working in radians.

22. **(E)**

$$\frac{x^2 - 5x + 6}{x - 2} = \frac{(x-3)(x-2)}{x-2} = x - 3 > 0$$

$$x > 3$$

23. **(A)**

If $\sin x = \dfrac{1}{2}$ then $\cos x = \dfrac{\sqrt{3}}{2}$, and if $\sin y = \dfrac{\sqrt{2}}{2}$, then $\cos y = \dfrac{\sqrt{2}}{2}$, as we are considering only positive values:

$$\tan(x + y) = \frac{\sin(x+y)}{\cos(x+y)}$$

$$\sin(x + y) = \sin x (\cos y) + \sin y (\cos x)$$

$\cos(x + y) = \cos x \,(\cos y) - \sin x (\sin y)$

$\sin(x + y) = \dfrac{1}{2} \times \dfrac{\sqrt{2}}{2} + \dfrac{\sqrt{2}}{2} \times \dfrac{\sqrt{3}}{2} = \dfrac{\sqrt{2}}{4} + \dfrac{\sqrt{6}}{4}$

$\cos(x + y) = \dfrac{\sqrt{3}}{2} \times \dfrac{\sqrt{2}}{2} - \dfrac{1}{2} \times \dfrac{\sqrt{2}}{2} = \dfrac{\sqrt{6}}{4} - \dfrac{\sqrt{2}}{4} = \dfrac{\sqrt{6} - \sqrt{2}}{4}$

$\tan(x + y) = \dfrac{\dfrac{\sqrt{2} + \sqrt{6}}{4}}{\dfrac{\sqrt{6} - \sqrt{2}}{4}} = \dfrac{(\sqrt{6} + \sqrt{2})}{(\sqrt{6} - \sqrt{2})} \dfrac{(\sqrt{6} + \sqrt{2})}{(\sqrt{6} + \sqrt{2})}$

$= \dfrac{(\sqrt{6})^2 + 2(\sqrt{6})(\sqrt{2}) + (\sqrt{2})^2}{6 - 2} = \dfrac{6 + 2\sqrt{6}\sqrt{2} + 2}{4}$

$= \dfrac{8 + 2\sqrt{3}\sqrt{2}\sqrt{2}}{4} = \dfrac{8 + 4\sqrt{3}}{4} = 2 + \sqrt{3}$

Calculator:

$x = \sin^{-1} \dfrac{1}{2} = (1 \div 2) \text{ INV sin} = 30°$

$y = \sin^{-1} \dfrac{\sqrt{2}}{2} = (2\sqrt{} \div 2) \text{ INV sin} = 45°$

$\tan(x + y) = (30° + 45°) \tan = 3.732$

Check that $2 + \sqrt{3} = 3.732$. Therefore (A) is correct.

24.   **(E)**

   From the definition of absolute value $|\,2x - 2\,|$ has to be either equal to $(2x - 2)$ or $-(2x - 2)$.

First case:

$2x - 2 > 0$

$|\,2x - 2\,| = 2x - 2$

$2x - 2 < 3$

$2x < 5$

$x < \dfrac{5}{2}$

Second case:

$2x - 2 < 0$

$$| \, 2x - 2 \, | = - 2x + 2$$

$$- 2x + 2 < 3$$

$$- 2x < 1$$

$$x > - \frac{1}{2}$$

If we represent the solutions graphically we will obtain the shaded area below:

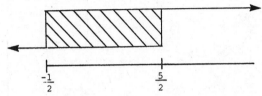

The solution set consists of $\{x \mid -\frac{1}{2} < x < \frac{5}{2}\}$.

25.　**(E)**

Since $\log a^x = x \log a$, thus

$$\log_{10}100^x = x \log_{10}100$$

Since $100 = 10^2$, we can apply the same rule, to get

$$x \log_{10}100 = x \log_{10}10^2 = 2x \log_{10}10 = 2x \ (\log_{10}10 = 1).$$

Therefore the expression is reduced to

$$\pm\sqrt{2x} = 2x$$

Squaring both sides

$$2x = 4x^2 \quad \text{or} \quad \frac{2x}{2x} = \frac{4x^2}{2x}, \quad \text{or} \quad 2x = 1, x = \frac{1}{2}$$

Calculator: The value of $\log_{10}100$ can be calculated on calculator as $100 \log = 2$.

26.　**(D)**

As x gets large, $2x^3$ gets large, so $\dfrac{1}{2x^3}$ gets small and tends to zero. So $1 +$

$\dfrac{1}{2x^3}$ tends to $1 + 0 = 1$.

27.　**(A)**

The lengths of the three sides can be calculated by the distance formula:

$$a = \sqrt{(5-1)^2 + (1-4)^2} = \sqrt{4^2 + 3^2} = \sqrt{25} = 5$$
$$b = \sqrt{(1-4)^2 + (4-0)^2} = \sqrt{3^2 + 4^2} = \sqrt{25} = 5$$
$$c = \sqrt{(4-5)^2 + (0-1)^2} = \sqrt{1^2 + 1^2} = \sqrt{2}$$

This is an isosceles triangle, whose area is $\frac{1}{2}ch$

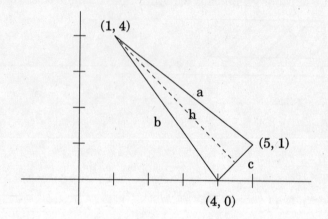

The value of h can be calculated by the Pythagorean Theorem, according to which

$$b^2 = h^2 + (\tfrac{1}{2}c)^2$$

or $\quad h^2 = b^2 - (\tfrac{1}{2}c)^2 = 5^2 - \frac{1}{4}(\sqrt{2})^2 = 25 - \frac{1}{2} = \frac{49}{2}$

or $\quad h = \sqrt{\dfrac{49}{2}} = \dfrac{7}{\sqrt{2}}$

The area is

$$A = \tfrac{1}{2}ch = \tfrac{1}{2}\sqrt{2} \times \frac{7}{\sqrt{2}} = \frac{7}{2}$$

Calculator: $\quad a = (5-1)\ \text{INV } x^2 + ((1-4)\ \text{INV } x^2) = \sqrt{\ } = 5$

$$b = (1-4)\ \text{INV } x^2 + (4\ \text{INV } x^2) = \sqrt{\ } = 5$$

$$c = (4-5)\ \text{INV } x^2 + (1 \pm \text{INV } x^2) = 2$$

28.   (C)
Applying the cosine formula:

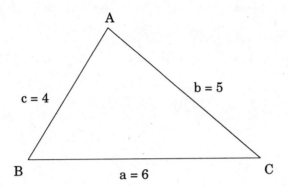

$$6^2 = 5^2 + 4^2 - 2(5)(4) \cos A$$

$$36 = 25 + 16 - 40 \cos A$$

$$36 - 41 = -40 \cos A$$

$$-5 = -40 \cos A$$

$$\cos A = \frac{-5}{-40} = \frac{1}{8}$$

$$A = \cos^{-1} \frac{1}{8} = 82.8°$$

Calculator: $36 - 25 - 16 = \div 40 \pm = $ INV COS $= 82.8$

29.   **(D)**

One root is a complex number, therefore the other root is the complex conjugate, i.e.,

$$\gamma_1 = 12 - 13i$$

$$\gamma_2 = 12 + 13i$$

Recall that the complex conjugate is obtained by replacing i by $-i$. The quadratic equations have the form

$$x^2 - (\gamma_1 + \gamma_2)x + \gamma_1 \gamma_2 = 0$$

$$\gamma_1 + \gamma_2 = 12 - 13i + 12 + 13i = 24$$

$$\gamma_1 \gamma_2 = (12 - 13i)(12 + 13i) = 12^2 + 13^2 = 144 + 169 = 313.$$

Therefore the quadratic equation is

$$x^2 - 24x + 313 = 0.$$

Calculator: 12 INV $x^2 + 13$ INV $x^2 = 313$

30.　**(C)**

The expression $(\sin x)\sec x + \cot x$ can be written as:

$$\sin x \frac{1}{\cos x} + \frac{\cos x}{\sin x} = \frac{\sin^2 x + \cos^2 x}{(\sin x)\cos x}$$

$$= \frac{1}{\dfrac{\sin 2x}{2}} = \frac{2}{\sin 2x}$$

31.　**(D)**

$$A^1 = (A*B)*(B*A) = \left(A^B\right)*\left(B^A\right) = \left(A^B\right)^{B^A} = A^{B^1 \cdot B^A} = A^{B(1+A)}.$$

Matching exponents, $B^{(1+A)} = 1 = B^\circ$ since any non-zero integer B raised to the zero power equals one. So, $(1+A) = 0$, which requires $A = -1$.

32.　**(E)**

The expression can be simplified according to the following rules:

$$\log_{10} x^y = y \log_{10} x, \quad \ln x^y = y \ln x$$

$$\log_{10} \frac{x}{y} = \log_{10} x - \log_{10} y, \quad \ln \frac{x}{y} = \ln x - \ln y$$

Therefore,

$$\log_{10} 10^{3.5} + \ln \frac{1}{5^{-2}} - \log_{10}\sqrt{25} - \ln\left(\frac{27}{30}\right)$$

$$= 3.5 \log_{10} 10 + 2 \ln 5 - \log_{10} 5 - \ln 27 + \ln 30$$

$$= 3.5 + 2(1.61) - (0.7) - 3.296 + 3.4$$

$$= 6.124$$

Calculator:　$(10 \log \times 3.5) + (5 \ln \times 2) - (5 \log) - (27 \ln) + (30 \ln)$

$$= 6.124$$

33. **(A)**

$\det A = (x-2)(x-3) - 0 = x^2 - 5x + 6$

$x^2 - 5x + 6 = 2x - 4$

$x^2 - 7x + 10 = 0.$

The roots are x_1 and x_2, where

$$x_1 = \frac{-(-7) + \sqrt{49 - 4 \times 1 \times 10}}{2}$$

$$= \frac{7+3}{2} = 5$$

and $\quad x_2 = \frac{-(-7) - \sqrt{49 - 4 \times 1 \times 10}}{2}$

$$= \frac{7-3}{2} = 2$$

Calculator: $\quad \sqrt{49 - 4 \times 1 \times 10} = 49 - (4 \times 1 \times 10) = \sqrt{} = 3$

$$x_1 = 7 + 3 = \div 2 = 5$$

$$x_2 = 7 - 3 = \div 2 = 2$$

34. **(D)**

The equation can be rewritten as follows:

$$\frac{(x-2)^2}{1} - \frac{(y-2)^2}{1} = 1$$

The center occurs at (2, 2) and:

$a^2 = 1 \qquad c = $ distance between focus and center

$b^2 = 1 \qquad c = \sqrt{a^2 + b^2} = \sqrt{2}$

The coordinates are given by:

$$F_1 = (2 - \sqrt{2}, 2) \text{ and } F_2 = (2 + \sqrt{2}, 2)$$

35. **(D)**

If $\sin^2 3x = 0.75$, then $\sin 3x = \pm\sqrt{0.75} = \pm 0.866$

If $\sin 3x = +0.866$, then $3x$ is in 1st or 2nd quadrant. Therefore

$$3x = \sin^{-1} 0.866 = 60° \text{ or } 120°$$

or $\quad x = \dfrac{60°}{3} = 20°$ or $x = \dfrac{120°}{3} = 40°$

If sin $3x = -0.866$, then $3x$ is in 3rd or 4th quadrant. Therefore

$$3x = -60° = 360 - 60 = 300° \text{ or } x = \dfrac{300}{3} = 100°$$

or $\quad 3x = -120° = 360 - 120 = 240° \text{ or } x = \dfrac{240}{3} = 80°$

Calculator: $\quad 0.75 \sqrt{} = 0.866$

$$.866 \text{ INV SIN} = 60$$

$$-0.866 \text{ INV SIN} = -60$$

36. **(B)**

If the roots satisfy $\dfrac{r_1}{r_2} = 3$ and their product is

$$\dfrac{c}{a} = \dfrac{12}{4} = 3$$

then: $\quad r_1 r_2 = 3 \qquad r_1 = 3r_2$

$$3r_2(r_2) = 3, \quad 3r_2{}^2 = 3$$

$$\dfrac{r_1}{r_2} = 3 \qquad r_2 = \pm 1 \text{ and } r_1 = \pm 3$$

Substituting into the equation:

If $r_1 = 3$: $\quad 4(3)^2 + q(3) + 12 = 0$

$$36 + 3q + 12 = 0$$

$$3q = -12 - 36$$

$$3q = -48$$

$$q = -16$$

If $r_1 = -3$: $\quad 4(-3)^2 + q(-3) + 12 = 0$

$$36 - 3q + 12 = 0$$

$$3q = 48, \ q = 16$$

If $r_2 = -1$: $4(1)^2 + q(1) + 12 = 0$

$$4 + q + 12 = 0$$

$$q = -16$$

If $r_2 = 1$: $4(-1)^2 + q(-1) + 12 = 0$

$$q = 16$$

We substitute the values of q obtained into $4x^2 + qz + 12$. We obtain:

$$4x^2 + 16x + 12 = x^2 + 4x + 3 \tag{1}$$

$$4x^2 - 16x + 12 = x^2 - 4x + 3 \tag{2}$$

Eq. (1) has roots -3 and -1 and Eq. (2) has roots 3 and 1, so 16 and -16 are answers.

37.   **(B)**

The general form of a term of the expansion of $(A + B)^n$ is given by:

$$\binom{n}{r} A^{n-r} B^r$$

$$\binom{4}{1} A^{4-1} B^1 = \frac{4!}{1!(4-1)!} \times (-4x)^3 \times (-3y)^1$$

$$= 4(-64x^3) \times (-3y) = 768x^3 y$$

Calculator: 4 INV nCr 1 = 4

$$4 \times (4 \pm \text{INV } x^y 3) \times 3 \pm = 768.$$

38.   **(B)**

If a statement is true, then its contrapositive is also true. Given that

"If Jack can run, then Jack can fly"

is true, we know that

"If Jack can't fly, then Jack can't run"

is true. We also know "Jack can't fly" so we can conclude that "Jack can't run."

39.   **(B)**

There are 5 members to be selected from a group of 15 people. Therefore the total number of possible ways is

$$\binom{15}{5} = 3003.$$

We have to select 3 men out of 6 men, so the number of possible ways is

$$\binom{6}{3} = 20.$$

Two women have to be selected from 9 women. The total number of possible ways is

$$\binom{9}{2} = 36.$$

Hence the desired probability

$$= \frac{20 \times 36}{3003} = \frac{240}{1001}$$

Calculator: 15 INV nCr 5 = 3003

6 INV nCr 3 = 20

9 INV nCr 2 = 36.

40. **(C)**

Let $y = \arcsin x$

$$f(x) = \sin 2y \quad = \quad 2 \sin y \cos y$$

$$= \quad 2 \sin(\arcsin x) \, [\cos(\arcsin s)]$$

$$= \quad 2x \, (\pm\sqrt{1-x^2})$$

So $\quad f\left(\dfrac{1}{3}\right) = 2\left(\dfrac{1}{3}\right) \times \sqrt{1-\left(\dfrac{1}{3}\right)^2} = \dfrac{2}{3} \times \left(\pm\sqrt{\dfrac{8}{9}}\right) = \dfrac{2}{3} \times \left(\dfrac{\pm 2\sqrt{2}}{3}\right) = \dfrac{\pm 4\sqrt{2}}{9}$

Since $\dfrac{1}{3}$ is a positive number, $\arcsin \dfrac{1}{3}$ will be in first and second quadrant.

In first quadrant $\arcsin \dfrac{1}{3} = 0.3398$ rad

In second quadrant $\arcsin \dfrac{1}{3} = \pi - 0.3398 = 2.802$, therefore

$$f(\dfrac{1}{3}) = \sin 2\,(0.3398) = 0.6285$$

and $\quad f(\dfrac{1}{3}) = \sin 2\,(2.802) = -0.6285$

Calculator: $1 \div 3 = $ INV SIN $= 0.3398 \times 2 = $ SIN $= 0.6285$

$\pi - 0.3398 = \times 2 = $ SIN $= -0.6285$

Check that (C) is the same as the above values.

41.   **(A)**
    Squaring both sides:

$$x^2 = 4x + 7$$

or $$x^2 - 4x - 7 = 0$$

$$x = \frac{-(-4) \pm \sqrt{(-4)^2 - 4 \times 1 \times (-7)}}{2}$$

Calculator: $x_1 = 4 \pm \text{INV } x^2 - (4 \times 1 \times 7 \pm) = \sqrt{\ } + 4 = \div 2 = 5.32$

$x_2 = 4 \pm \text{INV } x^2 - (4 \times 1 \times 7 \pm) = \sqrt{\ } \pm + 4 = \div 2 = -1.32$

42.   **(D)**
    The sum of the arithmetic progression with n terms, first term as a, and common difference as d, is given by

$$S = \frac{\pi}{2} \left[2a + (n - 1) d \right]$$

We have

$$n = 50, a = 2, \text{ and } d = 2$$

Thus

$$S = \frac{50}{2} \left[2(2) + (50 - 1)2 \right]$$

Calculator: $50 \div 2 \times ((2 \times 2) + (49 \times 2)) = 2550$

43.   **(E)**
    If

$$f(x) = \frac{x}{x+1},$$

then we need to find x expressed in terms of $f(x)$ to substitute it into $f(2x)$.

$$f(x) = \frac{x}{x+1}$$

$$[f(x)] (x + 1) = x$$

$$[f(x)] x + f(x) = x$$

$$x[f(x)] - x = -f(x)$$

$$x(f(x) - 1) = -f(x)$$

$$x = \frac{-f(x)}{(f(x) - 1)}$$

so $\qquad x = \dfrac{f(x)}{1 - f(x)}$ $\qquad\qquad\qquad\qquad\qquad$ (1)

$$f(2x) = \frac{2x}{2x + 1}$$

Substituting in for x from Eq. (1) we obtain:

$$f(2x) = \frac{2\left(\dfrac{f(x)}{1 - f(x)}\right)}{2\left(\dfrac{f(x)}{1 - f(x)}\right) + 1} = \frac{\dfrac{2f(x)}{1 - f(x)}}{\dfrac{2f(x)}{1 - f(x)} + \dfrac{1 - f(x)}{1 - f(x)}}$$

$$= \frac{2f(x)}{f(x) + 1}$$

44. **(C)**

The width will be multiplied by 3.5 and the length by 6.1, so the area will be $3.5 \times 6.3 = 22.05$ times the original area $= 22.05 \times 5 = 110.25$.

Calculator: $\quad 3.5 \times 6.3 \times 5 = 110.25$

45. **(D)**

Let $n = 0.\overline{32}$

$$3.\overline{23} = 10n$$

$$323.\overline{23} = 1000n$$

$$1000n - 10n = 323.\overline{23} - 3.\overline{23}$$

$$990n = 320$$

$$n = \frac{320}{990} = \frac{32}{99}$$

46. **(D)**

If $f(x) = \ln(x^2) = y$ then

$$ey = x^2$$

$$(e^y)^{\frac{1}{2}} = x$$

$$e^{\frac{y}{2}} = x$$

So $\qquad f^{-1}(x) = e^{\frac{x}{2}}.$

The y-intercept is where $x = 0$.

$$y = f^{-1}(\phi) = e^{\frac{\phi}{2}} = e^{\phi} = 1$$

So the y-intercept is (0, 1).

47. **(E)**

If $(\sin x) \cos x > 0$ then either $\sin x > 0$ and $\cos x > 0$ or $\sin x < 0$ and $\cos x < 0$. In either case

$$\tan x = \frac{\sin x}{\cos x}$$

will be greater than 0, and

$$\sec x \,(\csc x) = \frac{1}{\cos x} \times \frac{1}{\sin x}$$

will be greater than 0.

We know that $\sin x$ and $\cos x$ are both positive in the first quadrant, $0 < x < \dfrac{\pi}{2}$, and both negative in the third quadrant, $\pi < x < \dfrac{3\pi}{2}$. So all three statements are true.

48.   **(E)**

The area of the top square surface $= 15 \times 15 = 225$ cm^2.

The area of the top circle $= \pi \times \left(\dfrac{15}{2}\right)^2 = 56.25\pi.$

The shaded area at the top surface $= 225 - 56.25\pi$

Volume $=$ height $\times$ area at the top surface

$\qquad = 25\,(225 - 56.25\pi) = 1207$ cm^3

Calculator: $225 - (56.25 \times \pi) = \times 25 = 1207$ cm^3

49.   **(D)**

Call the arc AB. $AB = \pi r$ (given). If we multiply both sides by 2 we obtain:

$$2AB = 2\pi r = \pi d, \; AB = \frac{\pi}{2}d$$

$$\frac{AB}{d} = \frac{\pi}{2} \approx \frac{3.14}{2} = 1.57$$

Calculator: $\pi \div 2 = 1.57$

50.   **(C)**

We see that $x - y = \dfrac{2}{t}$ and $x + y = 2t$. So

$$(x - y)(x + y) = \frac{2}{t} \times 2t = 4$$

$$x^2 - y^2 = 4$$

$$\frac{x^2}{4} = \frac{y^2}{4} = 1 \text{ is a hyperbola.}$$

THE SAT SUBJECT TEST IN

Math
Level 2

PRACTICE TEST 6

SAT Mathematics
Level 2

Practice Test 6

Time: 1 Hour
       50 Questions

DIRECTIONS: Choose the best answer for each question and mark the letter of your selection on the corresponding answer sheet in the back of the book.

NOTES:

(1)   Some questions require the use of a calculator. You must decide when the use of your calculator will be helpful.

(2)   You may need to decide which mode your calculator should be in—radian or degree.

(3)   All figures are drawn to scale and lie in a plane unless otherwise stated.

(4)   The domain of any function f is the set of all real numbers x for which $f(x)$ is a real number, unless other information is provided.

REFERENCE INFORMATION: The following information may be helpful in answering some of the questions.

Volume of a right circular cone with radius r and height h	$V = \dfrac{1}{3}\pi r^2 h$
Lateral area of a right circular cone with circumference of the c and slant height l	$S = \dfrac{1}{2}cl$
Volume of a sphere with radius r	$V = \dfrac{4}{3}\pi r^3$
Surface Area of a sphere with r	$S = 4\pi r^2$
Volume of a pyramid with base area B and height h	$V = \dfrac{1}{3}Bh$

1. What is the domain of the function defined by

$$y = f(x) = \sqrt{-x+1} + 5?$$

(A) $\{ x \mid x \geq 0 \}$

(D) $\{ x \mid x \geq -1 \}$

(B) $\{ x \mid x \leq 1 \}$

(E) $\{ x \mid x \leq -1 \}$

(C) $\{ x \mid 0 \leq x \leq 1 \}$

2. What is the range of the function given in Problem 1?

(A) $\{ y \mid y \geq 5 \}$

(D) $\{ y \mid 0 < y \leq 5 \}$

(B) $\{ y \mid y > 5 \}$

(E) $\{$ all real numbers $\}$

(C) $\{ y \mid y > 0 \}$

3. In the given figure, $k =$

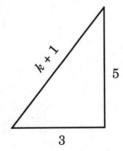

(A) 5.83

(D) -5.83

(B) 33

(E) 4.83

(C) 5.74

4. Given that $i = \sqrt{-1}$, $(-i)^{n-1} = 1$ if $n =$

(A) 2

(D) 5

(B) 3

(E) 6

(C) 4

5. If $f(x) = x^k + x^{2+k} - 10x$ and $f(3) = 4$, then $k =$

(A) 0

(D) 2.15

(B) 0.67

(E) 2.89

(C) 1.11

6. $\log_3 81 =$

(A) -4

(D) 4

(B) 2

(E) 1

(C) -2

7. Which of the following graphs could represent the equation
$y = ax^2 + bx + c$ where $b^2 - 4ac = 0$ and $a > 0$?

(A)

(D)

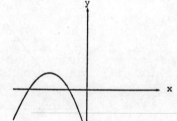

(B)

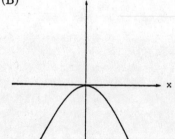

(E)

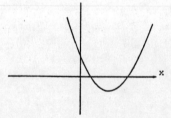

(C)

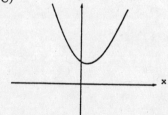

8. The base of a pyramid, in the shape of a square, has an area of 36 square centimeters. The side of the pyramid is 5 centimeters. What is the total surface area of the solid in square centimeters?

    (A) 84   (D) 132

    (B) 96   (E) 60

    (C) 72

9. An angle of measure $\dfrac{11\pi}{6}$ radians is equivalent to an angle measure of

    (A) 165°   (D) 330°

    (B) 300°   (E) 55°

    (C) 110°

10. The solution set of $\dfrac{x^2}{\ln x} > 0$ is

    (A) ϕ

    (B) $\{\, x \mid x > -1 \,\}$

    (C) $\{\, x \mid x > 0 \,\}$

    (D) $\{\, x \mid x > 1 \,\}$

    (E) $\{\, x \mid x \text{ is any real number} \}$

11. In the given figure, an ellipse inscribed in a rectangle has area A which is given by the formula $A = \pi ab$. If $A = 15\pi$ and $a = 5$, what is the perimeter of the rectangle?

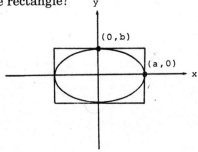

(A) 15   (D) 32

(B) 16   (E) 60

(C) 30

12. If $2 \sin 2x = 3 \cos 2x$, and $0 \leq 2x \leq \dfrac{\pi}{2}$, then $x =$

(A) 0   (D) $\dfrac{\pi}{3}$

(B) $\dfrac{1}{2}$ (E) $\dfrac{\pi}{2}$

(C) $\dfrac{\pi}{6}$

13. What is $\lim\limits_{x \to -5} \dfrac{2x^2 + 6x - 20}{x + 5}$?

(A) 0   (D) -6

(B) 6   (E) The limit does not exist.

(C) -14

14. If for all x, $f(x) = 3(a)^x$ and $f(x + 3) = 64f(x)$, then $a =$

(A) $\dfrac{2}{3}$ (D) $\dfrac{3}{4}$

(B) $\dfrac{8}{3}$ (E) $\dfrac{4}{3}$

(C) $\dfrac{3}{2}$

15. Which of the following could be the equation of the graph shown in the given figure?

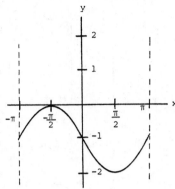

(A) $y = \sin x - 1$

(D) $y = \sin\left(\dfrac{\pi}{2} - x\right)$

(B) $y = \sin(-x)$

(E) $y = \sin\left(\dfrac{\pi}{2} - x\right) - 1$

(C) $y = \sin(-x) - 1$

16. At how many points does the graph of $y = x^4 - 2x^3 + x^2$ intersect the x-axis?

(A) 0

(D) 3

(B) 1

(E) 4

(C) 2

17. If $f(x) = 3x^3 - 2$ and $g(x) = \dfrac{1}{x}$, then $f^{-1}(g(2)) =$

(A) -1.625

(D) 15.625

(B) 0.94

(E) 0.794

(C) 22

18. If $f(x) = x^3$, and $f(a-1)^{\frac{4}{3}} = 3$, then $a =$

(A) 2.316

(D) 5.327

(B) 1.316

(E) 2.28

(C) 4.327

19. If a fair coin is tossed three times, what is the probability of getting exactly two heads?

(A) $\dfrac{1}{4}$

(D) 1

(B) $\dfrac{1}{8}$

(E) 0

(C) $\dfrac{3}{8}$

20. A geometric series has a first term $\dfrac{1}{2}$ and a common ratio of $\dfrac{1}{3}$. Find the sum of first 100 terms.

(A) 0.5

(D) 0.75

(B) 1

(E) 1.33

(C) 0.67

21. Which of the following figures represents the graph of the equation

$$4x^2 + 16y^2 = 64?$$

(A)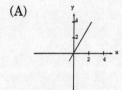

(B)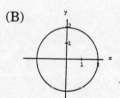

(C)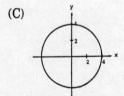

(D)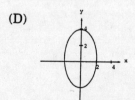

(E)

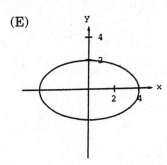

22. If a and b are both positive, odd integers, and $a > b$, then for which of the following functions will $f(a)$ be greater than $f(b)$?

   I. $f(x) = e^x$

   II. $f(x) = x^{-3}$

   III. $f(x) = (-5)^x$

(A)   I only.   (D)   I and III only.

(B)   II only.   (E)   I, II, and III.

(C)   III only.

23. Which quadrants of the plane contain points of the graph of $x^2 + y^2 < 1$?

(A)   First and second only.

(B)   First and third only.

(C)   Second and fourth only.

(D)   Third and fourth only.

(E)   First, second, third, and fourth.

24. If $f(x) = \dfrac{x-1}{x^3 - 3x^2 + 2x}$, for what value(s) of x is $f(x)$ undefined?

(A)   0   (D)   0, 1, and 2

(B)   0 and 2   (E)   0, -1, and -2

(C)   1 and 2

25. Let $f(x) = \log_5 \dfrac{3.5x - 1}{2}$, then $f^{-1}(5) =$

    (A) 893   (D) 210

    (B) 1786   (E) 1050

    (C) 6251

26. If $-\pi < x \le \pi$ and $\sin x = \sin 2x$, then $x =$

    (A) $\dfrac{\pi}{6}, \dfrac{5\pi}{6}$ (D) $\dfrac{\pi}{4}, \dfrac{3\pi}{4}$

    (B) $\dfrac{\pi}{3}, \dfrac{2\pi}{3}$ (E) $\dfrac{\pi}{2}, \pi$

    (C) $0, \pi$

27. If a straight line contains points (3, 4) and (−2, 7), then the equation of this straight line is

    (A) $y = -\dfrac{3}{5}x + \dfrac{27}{5}$ (D) $y = \dfrac{4}{3}x - \dfrac{7}{3}$

    (B) $y = \dfrac{11}{5}x - \dfrac{29}{5}$ (E) $y = -\dfrac{2}{7}x + \dfrac{29}{7}$

    (C) $y = -\dfrac{5}{3}x + \dfrac{29}{3}$

28. If $a = 2b$ and $c = 2d$, which of the following must be true? (Assume $c, d \ne 0$)

    I. $\dfrac{a}{c} = \dfrac{b}{d}$

    II. $ad = bc$

    III. $(ac)^2 = 2(bd)^2$

    (A) I only.   (D) I and III only.

    (B) II only.   (E) I, II, and III.

    (C) I and II only.

29. arccos (sin(arccot(sec $\frac{2\pi}{3}$))) =

    (A)   2.324   (D)   0.567

    (B) .1.108   (E)   3.521

    (C)   2.034

30. In $\triangle ABC$, $a = 2x$, $b = 3x + 2$, $c = 2$, and angle $C = 60°$.

Solve for x using the figure below.

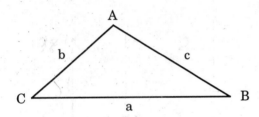

    (A) -0.368 (D)   0.368

    (B) -1.132 (E)   None of the above.

    (C)   1.132

31. If $f(x) = \sqrt{x}$, $g(x) = \frac{x-1}{4}$, and $h(x) = x^2$, what is $f(g(h(4)))$?

    (A)   2   (D)   1.94

    (B)   2.06   (E)   3.75

    (C)   2.24

32. If $f(x) = x^3 - x - 1$, then the set of all c for which $f(c) = f(-c)$ is

    (A)   All real numbers.   (D) $\{-1, 0, 1\}$

    (B) $\{0\}$ (E) ϕ

    (C) $\{0, 1\}$

33. What is the total surface area of a rectangular box, in square centimeters, with length 12 centimeters, width 6 centimeters, and height 15 centimeters?

(A)   342

(D)   432

(B)   864

(E)   1080

(C)   684

34. Let $f(x) = \dfrac{\sin x}{\cot x \cos x}$, then $f(0.955) =$

(A)   0.0003

(D) $\dfrac{1}{2}$

(B)   114.6

(E)   2

(C)   1.41

35. What type of figure does the graph of the set of pairs (x, y), where $x = \cos \theta$, $y = \sin \theta$, and $0 \le \theta < 2\pi$ turn out to be?

(A)   Cardioid

(D)   Hyperbola

(B)   Circle

(E)   Parabola

(C)   Ellipse

36. For all non-negative real numbers, a ≠ b, a * b is defined by the equation

$$a * b = \frac{2a^2 - b}{a - b}.$$

If $6 * x = 10 * 3$, then $x =$

(A) $\dfrac{35}{60}$

(D) $\dfrac{339}{95}$

(B) $\dfrac{177}{101}$

(E) $\dfrac{30}{59}$

(C) $\dfrac{354}{202}$

37. The side of a cube is equal to the diameter of the base of a cone. If the height of the cone is 5 centimeters and the volumes of the cube and the cone are equal, then the side of the cube is

(A)   1.14

(D)   2.62

(B)   1.618

(E)   5.24

(C)   1.31

38. The distance of the plane $2x - 3y + 5z + 3 = 0$ from the point $(2, 3, -1)$ is

(A) $\dfrac{7}{\sqrt{38}}$

(D) $\dfrac{21}{\sqrt{38}}$

(B) $-\dfrac{7}{\sqrt{38}}$

(E) $\dfrac{7}{\sqrt{20}}$

(C) $\dfrac{7}{38}$

39. The sum of the first 50 terms of an arithmetic series is 100. If the common difference is 2, what is the first term?

(A)   47

(D) -47

(B)   48

(E) -51

(C) -48

40. If $f(x) = (x - 1)^2 + (x + 1)^2$ for all real numbers x, which of the following are true?

I. $f(x) = f(-x)$

II. $f(x) = f(x + 1)$

III. $f(x) = |\, f(x)\, |$

(A)   None of these.

(D)   II and III only.

(B)   III only.

(E)   I, II, and III.

(C)   I and III only.

41. Evaluate c if $2^3 + 2^4 = (c-2)2^4$.

(A) $-\dfrac{1}{2}$ (D) 10

(B) 3 (E) $\dfrac{7}{2}$

(C) $\dfrac{5}{2}$

42. If $(x-7)$ divides $x^3 - 3k^3x^2 - 13x - 7$, then $k =$

(A) 1.7 (D) 1.2

(B) 4.63 (E) 1.3

(C) 5.04

43. For $0 < x < \dfrac{\pi}{4}$, $\sin x$ is not less than

(A) $\cos x$ (D) $\tan x$

(B) $\cos\left(\dfrac{\pi}{2} - x\right)$ (E) $\tan\left(\dfrac{\pi}{2} - x\right)$

(C) $\sin\left(\dfrac{\pi}{2} - x\right)$

44. Find the value of k if $e^{(5k-1)} \times 5^{(2-k)} = 10^k \times 7^{(k+1)}$.

(A) $\dfrac{17}{5}$ (D) $\dfrac{2}{15}$

(B) $\dfrac{5}{17}$ (E) $\dfrac{3}{17}$

(C) $\dfrac{6}{13}$

45. Which of the following functions of x are even functions?

(A) $\sin x$ (D) $\cot x$

(B) $\cos x$ (E) All of the above.

(C) $\tan x$

46. If $f(x) = -x^3 - 2x^2 + 4x - 8$, what is $f(-2x)$?

   (A) $8x^3 - 8x^2 - 8x - 8$ (D) $-8x^3 - 4x^2 - 8x - 8$

   (B) $-8x^3 + 8x^2 - 8x - 8$ (E) $8x^3 - 6x^2 - 6x - 8$

   (C) $8x^3 - 4x^2 - 8x - 8$

47. In the given figure, PQ is a diameter of the circle, R is a point on the circle, and PR = 3, QR = 4. What is the area of the circle?

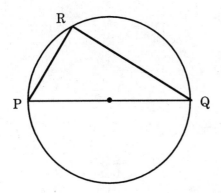

   (A) 25π (D) 5π

   (B) $\dfrac{7}{2}$ (E) $\dfrac{25}{4}\pi$

   (C) 10π

48. What is the contrapositive of the statement
       "If P is true, then Q is true"?

   (A) If P is false, then Q is true.

   (B) If P is true, then Q is false.

   (C) If P is false, then Q is false.

   (D) If Q is true, then P is true.

   (E) If Q is false, then P is false.

49. Using an ordinary deck of 52 playing cards, what is the probability of drawing three black cards in a row if each drawn card is not returned to the deck?

(A) $\dfrac{1}{8}$

(D) $\dfrac{4}{33}$

(B) $\dfrac{2}{17}$

(E) $\dfrac{5}{41}$

(C) $\dfrac{3}{25}$

50. $\dfrac{(n-2)!}{n!} - \dfrac{(n-1)(n-2)!}{n(n-1)!(n-1)} =$

(A) $\dfrac{1-n}{n(n-1)}$

(D) $\dfrac{n-2}{n}$

(B) 0

(E) $\dfrac{-n}{1-n}$

(C) $\dfrac{n^2-4n+4}{n(n-1)}$

SAT Mathematics Level 2

Practice Test 6
ANSWER KEY

1.	**(B)**	14.	**(E)**	27.	**(A)**	40.	**(C)**
2.	**(A)**	15.	**(C)**	28.	**(C)**	41.	**(E)**
3.	**(E)**	16.	**(C)**	29.	**(C)**	42.	**(D)**
4.	**(D)**	17.	**(B)**	30.	**(E)**	43.	**(B)**
5.	**(C)**	18.	**(A)**	31.	**(D)**	44.	**(B)**
6.	**(D)**	19.	**(C)**	32.	**(D)**	45.	**(B)**
7.	**(A)**	20.	**(D)**	33.	**(C)**	46.	**(A)**
8.	**(A)**	21.	**(E)**	34.	**(E)**	47.	**(E)**
9.	**(D)**	22.	**(A)**	35.	**(B)**	48.	**(E)**
10.	**(D)**	23.	**(E)**	36.	**(D)**	49.	**(B)**
11.	**(D)**	24.	**(B)**	37.	**(C)**	50.	**(B)**
12.	**(B)**	25.	**(B)**	38.	**(A)**		
13.	**(C)**	26.	**(B)**	39.	**(D)**		

DETAILED EXPLANATIONS
OF ANSWERS

1. **(B)**
   The only restriction for the domain is that $-x+1$ must be greater than or equal to zero.

   $$-x + 1 \geq 0$$

   $$\Rightarrow 1 \geq x$$

2. **(A)**

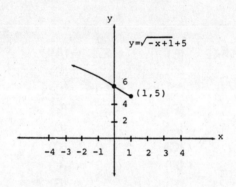

   Since $\sqrt{-x+1} \geq 0$, the range is all values of $y \geq 5$.

3. **(E)**
   The given figure is a right triangle. Therefore,

   $$5^2 + 3^2 = (k + 1)^2$$

   $$25 + 9 = k^2 + 2k + 1$$

   $$34 = k^2 + 2k + 1$$

   or $\quad k^2 + 2k - 33 = 0$

   Applying the quadratic formula:

   $$k = \frac{-2 \pm \sqrt{2^2 - 4 \times 1 \times (-33)}}{2}$$

Calculator: $\sqrt{2^2 - 4 \times 1 \times (-33)}$ = 2 INV $x^2 - (4 \times 33\pm) = \sqrt{} = 11.66$

$$k_1 = -2 + 11.66 = \div 2 = 4.83$$

$$k_2 = -2 - 11.66 = \div 2 = -6.83$$

Since $k + 1 = 4.83 + 1 = 5.83$ or $k + 1 = -6.83 + 1 = -5.83$, and $k + 1$ cannot be negative, therefore $k + 1 = 5.83$, and $k = 4.83$.

4. **(D)**

$$(-i)^1 = -i$$

$$(-i)^2 = -1$$

$$(-i)^3 = -i$$

$$(-i)^4 = 1$$

Therefore, $n - 1 = 4 \Rightarrow n = 5$

Note: i or $-i$ raised to any power which is a multiple of 4 will be equal to 1.

5. **(C)**

$$f(3) = 3^k + 3^{2+k} - 30 = 4$$

$$3^k + 3^2 \times 3^k = 34$$

$$3^k + 9 \times 3^k = 34$$

$$3^k (1 + 9) = 34$$

$$10 \times 3^k = 34$$

$$3^k = 3.4$$

Taking logs,

$$k \times \log 3 = \log 3.4$$

$$k = \log 3.4 \div \log 3 \approx 1.11$$

6.   **(D)**

$$\log_3 81 = \log_3 3^4 = 4 \log_3 3 = 4 \times 1 = 4.$$

We can find this value on the calculator by using the following formula:

$$\log_x y = \frac{\log_{10} y}{\log_{10} x} = \frac{\ln y}{\ln x}$$

Therefore, $\log_3 81 = \dfrac{\log_{10} 81}{\log_{10} 3} = \dfrac{1.908}{0.477} = 4$

Calculator: 81 log ÷ 3 log = 4

or   81 ln ÷ 3 ln = 4

7.   **(A)**

If $b^2 - 4ac = 0$, then there is only 1 value for x for which $y = 0$. Therefore, the parabola is tangent to the x-axis. Whenever a > 0, the curve is concave up. Only the figure of choice (A) satisfies these two conditions.

8.   **(A)**

If the area of the square base = 36, then each side of the square is 6. The figure of the pyramid opened up is shown below.

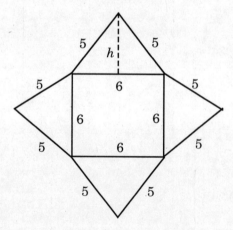

$$\text{height} = h = \sqrt{5^2 - 3^2} = \sqrt{25 - 9} = \sqrt{16} = 4$$

T = Total surface area = (Area of square) + 4 (Area of one triangle)

$$= 6^2 + 4(\frac{1}{2}) \times 4 \times 6 = 84$$

Calculator: $6 \times 4 \div 2 \times 4 = + 6$ INV $x^2 = 84$

9.   **(D)**
   An angle of measure π radians is equivalent to an angle measure of 180°. Therefore,

$$\frac{11\pi}{6} = \frac{11(180°)}{6} = 330$$

Calculator: $11 \times 180 \div 6 = 330$

10.   **(D)**
   The natural log of x, ln x, is undefined for $x \leq 0$ and is negative for $0 < x < 1$.

$\dfrac{x^2}{\ln x}$ is undefined for $x = 1$. Therefore, the solution set is $\{ x \mid x > 1 \}$.

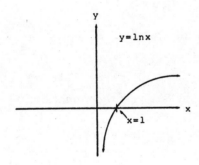

11.   **(D)**
   If $A = 15\pi$ and $a = 5$, then $b = 3$. This means that two of the sides of the rectangle are $2a = 10$ units each and the other two sides are $2b = 6$ units each.

   Perimeter = 2(10) 2(6) = 32

12.   **(B)**
   Applying the trigonometric identity:

$$\sin^2 x + \cos^2 x = 1$$

or $\cos x = \sqrt{1 - \sin^2 x}$

$$2 \sin 2x = \sqrt{1 - \sin^2 2x}$$

Squaring both sides:

$$4 \sin^2 2x = 9(1 - \sin^2 2x)$$

or $\quad 4 \sin^2 2x = 9 - 9 \sin^2 2x$

$$13 \sin^2 2x = 9$$

$$\sin 2x = \pm \sqrt{\frac{9}{13}}$$

Since $\quad 0 \leq 2x \leq \dfrac{\pi}{2}$

$$\sin 2x = + \sqrt{\frac{9}{13}}$$

$$2x = \arcsin \sqrt{\frac{9}{13}}$$

$$x = \frac{1}{2} \arcsin \sqrt{\frac{9}{13}}$$

Calculator: $\quad 9 + 13 = \sqrt{}$ INV SIN $\div 2 = 0.5$

Be sure that you are working in radians.

13.   **(C)**
The limit can be found by factoring the numerator.

$$\lim_{x \to -5} \frac{2x^2 + 6x - 20}{x + 5} = \lim_{x \to -5} \frac{(2x - 4)(x + 5)}{x + 5}$$

$$= \lim_{x \to -5} 2x - 4 = 2(-5) - 4 = -14$$

14.   **(E)**

$$f(x) = (3a)^x$$

$$f(x + 3) = (3a)^{x + 3} = 64(3a)^x$$

or $\quad (3a)^3 = 64$

$$(3a)^3 = (4)^3$$

$$3a = 4$$

$$a = \frac{4}{3}$$

With the help of calculator, we have

$$(3a) = \sqrt[3]{64}$$

Calculator: 64 INV x^y $\overset{\frac{1}{3}}{3} = 4$

15.   **(C)**
   The given figure is the graph of $y = \sin(-x) - 1$. The other graphs are:

(A)

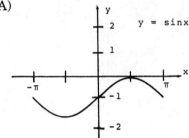

(B)

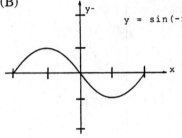

(D)

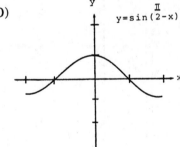

(E)

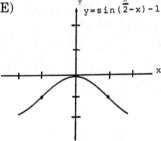

16.   **(C)**
   When a graph intersects the x-axis, it means that $y = 0$. Setting $y = 0$ in the equation:

$$y = x^4 - 2x^3 + x^2, \text{ we get}$$

$$x^4 - 2x^3 + x^2 = 0$$

$$x^2(x^2 - 2x + 1) = 0$$

$$x^2(x-1)^2 = 0$$

$$x = 0, \quad x = 1.$$

The curve intersects the x-axis at $x = 0$ and $x = 1$.

Note: The roots $x = 0$ and $x = 1$ both have a multiplicity of 2.

17.   **(B)**

$$g(2) = \frac{1}{2} = 0.5$$

To find $f^{-1}(x)$, let

$$y = 3x^3 - 2.$$

Replace x by y and y by x:

$$x = 3y^3 - 2$$

$$3y^3 = x + 2$$

$$y^3 = \frac{1}{3}(x + 2)$$

$$y = \sqrt[3]{\frac{1}{3}(x+2)}$$

or $$f^{-1}(x) = \sqrt[3]{\frac{1}{3}(x+2)}$$

$$f^{-1}(g(2)) = f^{-1}(0.5) = \sqrt[3]{\frac{1}{3}(0.5+2)}$$

Calculator: $0.5 + 2 = \div 3 = \text{INV } x^y 3 = 0.94$

18.   **(A)**

$$f(a-1)^{\frac{4}{3}} = [(a-1)^{\frac{4}{3}}]^3 = (a-1)^{\frac{4}{3}x^3} = (a-1)^4 = 3$$

$$(a-1) = (3)^{\frac{1}{4}}$$

$$a = 1 + (3)^{\frac{1}{4}} = 2.316$$

Calculator: $3 \text{ INV } x^y 4 = + 1 = 2.316$

19. **(C)**

There are 3 different possibilities in which we can get two heads in three tosses:

HHT, HTH, THH .

Therefore the total probability is

$$\left(\frac{1}{2}\times\frac{1}{2}\times\frac{1}{2}\right)+\left(\frac{1}{2}\times\frac{1}{2}\times\frac{1}{2}\right)+\left(\frac{1}{2}\times\frac{1}{2}\times\frac{1}{2}\right)$$

$$=\frac{1}{8}+\frac{1}{8}+\frac{1}{8}=\frac{3}{8}.$$

With a calculator, we see that there are $\binom{3}{2}$ possible ways in which we can get two heads in three tosses, and each outcome has a probability

$$\frac{1}{2}\times\frac{1}{2}\times\frac{1}{2}$$

Therefore, total probability is $\binom{3}{2}\times\frac{1}{8}$.

Calculator: 3 INC nCr2 = 3

20. **(D)**

The sum of the first n terms in a geometric series is

$$S_n - \frac{t_1(1-r^n)}{1-r}$$

where t_1 is the first term, r is the common ratio. Thus

$$S_{100}=\frac{\frac{1}{2}\left(1-\left(\frac{1}{3}\right)^{100}\right)}{1-\frac{1}{3}}\approx\frac{\frac{1}{2}}{1-\frac{1}{3}}=0.75$$

Calculator: $1 \div 2 \div (1 - (1 \div 3)) = 0.75$

21. **(E)**

$$4x^2 + 16y^2 = 64$$

can be rewritten as

$$\frac{x^2}{16}+\frac{y^2}{4}=1.$$

This is an equation of an ellipse with center at the origin. The graph intersects the x-axis at $x = 4$ and $x = -4$. It intersects the y-axis at $y = 2$ and $y = -2$.

22.   **(A)**

Take two arbitrary, positive, odd integers for our analysis. The use of these numbers will provide counterexamples for II and III.

I.   Since e^x is an increasing function (grows larger as x gets larger), we see that

$$a > b \Rightarrow e^a > e^b, \text{ so } f(a) > f(b).$$

II. $f(3) = 3^{-3} = \dfrac{1}{27}$ and $f(1) = 1^{-3} = 1$, letting $a = 3, b = 1$.

$$\dfrac{1}{27} < 1 \Rightarrow \text{No}, f(a) \not> f(b).$$

III. $f(3) = (-5)^3 = -125$ and $f(1) = (-5)^1 = -5$, letting $a = 3, b = 1$.

$$-125 < -5 \Rightarrow \text{No}, f(a) \not> f(b).$$

23.   **(E)**

The solution set of $x^2 + y^2 < 1$ is everything in the shaded area and it contains points from all four quadrants.

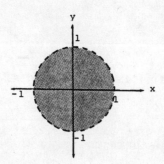

24.   **(B)**

$$f(x) = \frac{x-1}{x^3 - 3x^2 + 2} = \frac{x-1}{x(x^2 - 3x + 2)} = \frac{x-1}{x(x-1)(x-2)} = \frac{1}{x(x-2)}$$

$f(x)$ is undefined when $x(x - 2) = 0$, or $x = 0$ and $x = 2$.

The factors of $x^2 - 3x + 2$ can be calculated by using the quadratic formula:

$$x = \frac{-(-3) \pm \sqrt{9 - 4 \times 1 \times 2}}{2} = \frac{3 \pm 1}{2}$$

Calculator: $(3 + 1) \div 2 = 2$

$$(3 - 1) \div 2 = 1$$

Therefore, the factors are $(x - 2)$ and $(x - 1)$

25. **(B)**

Let $y = \log_5 \dfrac{3.5x-1}{2}$

Replacing x by y, and y by x:

$$x = \log_5 \dfrac{3.5y-1}{2}$$

or $\quad 5^x = \dfrac{3.5y-1}{2}$

or $\quad 2 \times 5^x = 3.5y - 1$

or $\quad y = \dfrac{2 \times 5^x + 1}{3.5}$

Thus $\quad f^{-1}(x) = \dfrac{2 \times 5^x + 1}{3.5}$

$$f(5) = \dfrac{2 \times 5^5 + 1}{3.5} = 1786$$

Calculator: $\quad 2 \times (5 \text{ INV } x^y 5) + 1 = \div 3.5 = 1786$

26. **(B)**

Since $\sin 2x = 2 \sin x \cos x$,

$\quad \sin x = \sin 2x$ implies $\sin x = 2 \sin x \cos x$

or $\quad 1 = 2 \cos x$

or $\quad \cos x = \dfrac{1}{2}$

$\quad x = \arccos \dfrac{1}{2} = \dfrac{\pi}{3}, \ \pi - \dfrac{\pi}{3} = \dfrac{2\pi}{3}.$

27. **(A)**

The equation of a straight line with slope m and y-intercept c is given by

$$y = mx + c$$

slope $\quad m = \dfrac{7-4}{-2-3} = \dfrac{3}{-5} = -\dfrac{3}{5}$

Thus $\quad y = -\dfrac{3}{5} x + c$

Since points (3, 4) and (– 2, 7) satisfy this equation,

$$3 = -\dfrac{3}{5}(4) + c$$

$$c = 3 + \dfrac{12}{5} = \dfrac{27}{5}$$

Thus the equation of the line is

$$y = -\dfrac{3}{5} x + \dfrac{27}{5}$$

28. **(C)**
Substitute $a = 2b$ and $c = 2d$ into each equation.

I. $\quad \dfrac{a}{c} = \dfrac{b}{d} \Rightarrow \dfrac{2b}{2d} = \dfrac{b}{d} \Rightarrow \dfrac{b}{d} = \dfrac{b}{d} \therefore$ True

II. $\quad ad = bc \Rightarrow (2b)d = b(2d) \Rightarrow 2bd = 2bd \therefore$ True

III. $\quad (ac)^2 = 2(bd)^2 \Rightarrow [(2b)(2d)]^2 = 2(bd)^2 \Rightarrow 4bd = 2bd \therefore$ False

29. **(C)**

$$\sec\dfrac{2\pi}{3} = \dfrac{1}{\cos\dfrac{2\pi}{3}}$$

Calculator: $\quad 2 \times \pi \div 3 = \cos \text{ INV} \dfrac{1}{x} = -2$

let $\quad$ arccot$(-2) = x$

then $\quad \cot x = -2$

$$\dfrac{1}{\tan x} = -2$$

or $\quad \tan x = -\dfrac{1}{2}$

$$x = \arctan\left(-\dfrac{1}{2}\right) = -0.464$$

$$1 \div 2 = \pm \text{ INV } \tan = -0.464$$

$$\sin(-0.464) = -0.447$$

$$0.464 \pm \sin = -0.447$$

$$\text{arccos}\,(-\,0.447) = 2.034$$

$$0.447 \pm \text{INV COS} = 2.034$$

30.　**(E)**

Applying the cosine formula:

$$c^2 = a^2 + b^2 - 2ab\cos C$$

$$2^2 = (2x)^2 + (3x+2)^2 - 2(2x)(3x+2)\cos 60°$$

$$4 = 4x^2 + 9x^2 + 12x + 4 - (12x^2 + 8x)\frac{1}{2}$$

$$4 = 13x^2 + 12x + 4 - 6x^2 - 4x$$

$$4 = 7x^2 + x + 4$$

or　　$7x^2 + 8x = 0$

or　　$x(7x+8) = 0$

The only possible solutions are $x = 0$ and $x = -\dfrac{8}{7} = -1.143$. Since neither choice is given in the list of possibilities, the correct answer is "None of the above."

31.　**(D)**

$$h(4) = 4^2 = 16$$

$$g(h(4)) = \frac{16-1}{4} = \frac{15}{4} = 3.75$$

$$f(g(h(4))) = \sqrt{3.75} = 1.94$$

32.   **(D)**

$$f(c) = c^3 - c - 1, \quad f(-c) = -c^3 + c - 1$$

$$c^3 - c - 1 = -c^3 + c - 1$$

$$c^3 = c$$

∴ $c = -1, 0, \text{ or } 1$

33.   **(C)**

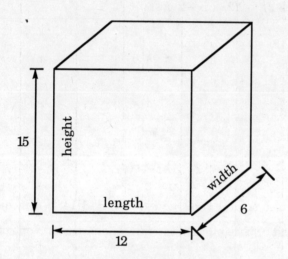

Total Surface Area = sum of the surface areas of the six sides

= 2(length × width) + 2(width × height) + 2(height × length)

= 2(12 × 6) + 2(6 × 15) + 2(15 × 12)

= 684

34.   **(E)**

$$f(x) = \frac{\sin x}{\cot x \cos x} = \frac{\sin x}{\left(\dfrac{1}{\tan}x\right)\cos x} = \frac{\tan x \sin x}{\cos x}$$

$$f(0.955) = \frac{\tan(0.955)\sin(0.955)}{\cos(0.955)}$$

Calculator: 0.955 tan × 0.9555 sin = ÷ 0.955 cos = 1.997 ≈ 2.

The calculation is easier if we recognize that

$$f(x) = \tan x \frac{\sin x}{\cos x} = \tan x \tan x = \tan^2 x .$$

Calculator: 0.9555 tan x INV x^2 = 1.997 ≈ 2

35.   **(B)**

$$x = \cos \theta$$

$$y = \sin \theta$$

$$x^2 + y^2 = \cos^2 \theta + \sin^2 \theta = 1$$

is a circle centered at (0, 0) with radius 1.

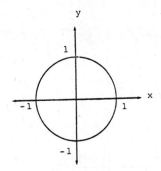

36.   **(D)**

$$6*x = \frac{2(6^2) - x}{6 - x} = \frac{72 - x}{6 - x}$$

$$10*3 = \frac{2(10^2) - 3}{10 - 3} = \frac{2(100) - 3}{10 - 3} = \frac{197}{7}$$

Therefore:

$$\frac{72 - x}{6 - x} = \frac{197}{7}$$

or $7(72 - x) = (6 - x)197$

$$504 - 7x = 1182 - 197x$$

$$190x = 678$$

$$x = \frac{678}{190} = \frac{339}{95}$$

37.   (C)
The cube with side r has the volume r^3. The cone has volume

$$V = \frac{1}{3} \times \pi \times \left(\frac{r}{2}\right)^2 \times 5 = \frac{5}{12}\pi r^2.$$

Since the volume of the cube is equal to the volume of the cone,

$$r^3 = \frac{5}{12}\pi r^2$$

or $r = \frac{5}{12}\pi = 1.31$ cm

Calculator: $5 \div 12 \times \pi = 1.3089 \approx 1.31$

38.   (A)
The distance of the plane $Ax + By + Cz + D = 0$, from the point (x_1, y_1, z_1) is given by

$$\text{Distance} = \frac{|Ax_1 + By_1 + Cz_1 + D|}{\sqrt{A^2 + B^2 + C^2}}$$

We have $A = 2, B = -3, C = 5, D = 3, x_1 = 2, y_1 = 3, z_1 = -1$. Using these values:

$$\text{Distance} = \frac{|2(2) + (-3)(3) + 5(-1) + 3|}{\sqrt{2^2 + (-3)^2 + 5^2}}$$

$$= \frac{|4 - 9 - 5 + 3|}{\sqrt{4 + 9 + 25}} = \frac{7}{\sqrt{38}}$$

The numbers in the distance formula can be simplified by the use of a calculator.

Calculator: $2 \times 2 + (3 \pm \times 3) + (5 \times 1 \pm) + 3 = -7$ or $|-7| = 7$

$$2 \text{ INV } x^2 + 3 \pm \text{ INV } x^2 = + 5 \text{ INV } x^2 = 38$$

39.   (D)
The sum of n terms in arithmetic series with first term t_1, and common difference d is given by

$$S_n = \frac{n}{2}[2t_1 + (n-1)d]$$

In this problem $S_n = 100, n = 50, d = 2$, thus,

$$100 = \frac{50}{2}(2t_1 + (50 - 1)2)$$

$$100 = 25(2t_1 + 98)$$

$$\frac{100}{25} = 2t_1 + 98$$

$$4 = 2t_1 + 98$$

$$2t_1 = 4 - 98 = -94$$

$$t_1 = -\frac{94}{2} = -47$$

The values of $(50 - 1)2$, $\frac{94}{2}$ can be calculated by using the calculator.

Calculator: $(50 - 1) \times 2 = 98$

$$94 \div 2 = 47$$

40.   **(C)**

I.   True: $f(-x)$ $= (-x - 1)^2 + (-x + 1)^2 = [-(x + 1)]^2 + [-(x - 1)]^2$

$= (x + 1)^2 + (x - 1)^2 = f(x)$

II.   False: $f(x + 1) = (x + 1 - 1)^2 + (x + 1 + 1)^2$

$= x^2 + (x + 2)^2 \neq f(x)$

III.   True: $|f(x)|$ $= |(x - 1)^2 + (x + 1)^2 |$

$= (x - 1)^2 + (x + 1)^2 = f(x)$

41.   **(E)**

$$2^3 + 2^4 = (c - 2)2^4$$

$$8 + 16 = (c - 2)16$$

$$24 = (c - 2)16$$

$$\frac{24}{16} = c - 2$$

$$c = \frac{24}{16} + 2 = \frac{7}{2}$$

Calculator: 2 INV 3 + 2 INV 4 = 24

$$24a\frac{b}{c}16 + 2 = 3.5 = \frac{3 \times 2 + 1}{2} = \frac{7}{2}$$

42. **(D)**

Using long division:

$$
\begin{array}{r}
x^2 + (7 - 3k^3)x + (36 - 21k^3) \\
x - 7 \overline{\smash{\big)}\ x^3 + 3k^3 x^2 - 13x - 7} \\
\underline{x^3 - 7x^2} \\
(7 - 3k^3)x^2 - 13x \\
\underline{(7 - 3k^3)x^2 + (-49 + 21k^3)x} \\
(36 - 21k^3)x - 7 \\
\underline{(36 - 21k^3)x - 7(36 - 21k^3)} \\
0
\end{array}
$$

So $\quad 36 - 21k^3 = 1 \Rightarrow 35 = 21k^3 \Rightarrow k = \sqrt[3]{\dfrac{35}{21}}$

Calculator: $\quad 35 \div 21 = \text{INV } x^y\, 3 = 1.1856 \approx 1.2$

43. **(B)**

By inspection, it can be determined that $\sin x$ is less than $\cos x$, $\sin\left(\dfrac{\pi}{2} - x\right)$, $\tan x$, and $\tan\left(\dfrac{\pi}{2} - x\right)$ for $0 < x < \dfrac{\pi}{4}$. $\sin x = \cos\left(\dfrac{\pi}{2} - x\right)$ for all values of x.

44. **(B)**

Taking logarithm on both sides

$\ln\left[e^{(5k-1)} + 5^{(2-k)}\right] = \ln\left[10^k 7^{k+1}\right]$

$\ln e^{5k-1} + \ln 5^{2-k} = \ln 10^k + \ln 7^{k+1}$

$(5k - 1)\ln e + (2 - k)\ln 5 = k \ln 10 + (k + 1)\ln 7$

$(5k - 1) + (2 - k)(1.6) = 2.3k + (k + 1)1.95$

$5k - 1 + 3.2 - 1.6k = 2.3k + 1.95k + 1.95$

$0.85k = 0.25$

$k = \dfrac{0.25}{0.85} = \dfrac{25}{85} = \dfrac{5}{17}$

Calculator: $5 \ln = 1.6$

$10 \ln = 2.3$

$7 \ln = 1.95$

To simplify the last equation, we can use the following:

$2.3 + 1.95 + 1.6 - 5 = 0.85$

$3.2 - 1 - 1.95 = 0.25$

$25 \ a\frac{b}{c} \ 85 = \frac{5}{17}$

45.   **(B)**

For a function to be even, $f(x)$ must equal $f(-x)$ for all x.

$\sin(-x) = -\sin(x) \Rightarrow$ No

$\cos(-x) = \cos(x) \Rightarrow$ Yes

$\tan(-x) = \dfrac{\sin(-x)}{\cos(-x)} = \dfrac{-\sin x}{\cos x} = -\tan x \Rightarrow$ No

$\cot(-x) = \dfrac{\cos(-x)}{\sin(-x)} = \dfrac{\cos x}{-\sin x} = -\cot x \Rightarrow$ No

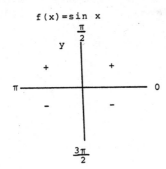

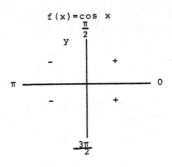

46.   **(A)**

$$f(-2x) = -(-2x)^3 - 2(-2x)^2 + 4(-2x) - 8$$

$$= -(-8x^3) - 2(4x^2) - 8x - 8$$

$$= 8x^3 - 8x^2 - 8x - 8$$

47.   **(E)**

Since the hypotenuse of the triangle is the diameter of the circle, $\triangle PQR$ is a right angle triangle.

$$(\text{Diameter})^2 = (\text{PR})^2 + (\text{RQ})^2$$

$$= 3^2 + 4^2 = 9 + 16 = 25$$

$\Rightarrow$ $\qquad$ Diameter $= \sqrt{25} = 5$

$\qquad\qquad$ Radius $= \dfrac{5}{2}$

$\qquad$ Area of the Circle $= \pi\left(\dfrac{5}{2}\right)^2 = \dfrac{25}{4}\pi$

Calculator: $\quad$ 3 INV x^2 + 4 INV $x^2 = \sqrt{} = 5$

48.   **(E)**

If the original logic statement is "If P, then Q" or $P \to Q$ symbolically, then we have the following:

(1)   The converse statement is $\sim P \to \sim Q$, as in choice (C).

(2)   The inverse statement is $Q \to P$, as in choice (D).

(3)   The contrapositive statement is $\sim Q \to \sim P$.

Notes: The original statements and the contrapositive are logically equivalent. The converse and the inverse are logically equivalent. $\sim P$ means "not P." So if P is true, $\sim P$ is false, and vice versa.

49.   **(B)**

$$\text{Pr(black card on first draw)} = \frac{26}{52}$$

$$\text{Pr(black card on second draw)} = \frac{25}{51}$$

$$\text{Pr(black card on third draw)} = \frac{24}{50}$$

$$\text{Pr(drawing three straight black cards)} = \frac{26}{52} \times \frac{25}{51} \times \frac{24}{50} = \frac{2}{17}$$

Calculator: $\quad 26a\dfrac{b}{c}52 \times 25a\dfrac{b}{c}51 \times 24a\dfrac{b}{c}50 = \dfrac{2}{17}$

50.   **(B)**

$$\frac{(n-2)!}{n!} - \frac{(n-1)(n-2)!}{n(n-1)!(n-1)} = \frac{(n-2)!}{n(n-1)(n-2)!} - \frac{(n-1)!}{n(n-1)(n-1)!}$$

$$= \frac{1}{n(n-1)} - \frac{1}{n(n-1)} = 0$$

THE SAT SUBJECT TEST IN

Math
Level 2

ANSWER SHEETS

SAT Math Level 2

Practice Test 1

1. Ⓐ Ⓑ Ⓒ Ⓓ Ⓔ
2. Ⓐ Ⓑ Ⓒ Ⓓ Ⓔ
3. Ⓐ Ⓑ Ⓒ Ⓓ Ⓔ
4. Ⓐ Ⓑ Ⓒ Ⓓ Ⓔ
5. Ⓐ Ⓑ Ⓒ Ⓓ Ⓔ
6. Ⓐ Ⓑ Ⓒ Ⓓ Ⓔ
7. Ⓐ Ⓑ Ⓒ Ⓓ Ⓔ
8. Ⓐ Ⓑ Ⓒ Ⓓ Ⓔ
9. Ⓐ Ⓑ Ⓒ Ⓓ Ⓔ
10. Ⓐ Ⓑ Ⓒ Ⓓ Ⓔ
11. Ⓐ Ⓑ Ⓒ Ⓓ Ⓔ
12. Ⓐ Ⓑ Ⓒ Ⓓ Ⓔ
13. Ⓐ Ⓑ Ⓒ Ⓓ Ⓔ
14. Ⓐ Ⓑ Ⓒ Ⓓ Ⓔ
15. Ⓐ Ⓑ Ⓒ Ⓓ Ⓔ
16. Ⓐ Ⓑ Ⓒ Ⓓ Ⓔ
17. Ⓐ Ⓑ Ⓒ Ⓓ Ⓔ
18. Ⓐ Ⓑ Ⓒ Ⓓ Ⓔ
19. Ⓐ Ⓑ Ⓒ Ⓓ Ⓔ
20. Ⓐ Ⓑ Ⓒ Ⓓ Ⓔ
21. Ⓐ Ⓑ Ⓒ Ⓓ Ⓔ
22. Ⓐ Ⓑ Ⓒ Ⓓ Ⓔ
23. Ⓐ Ⓑ Ⓒ Ⓓ Ⓔ
24. Ⓐ Ⓑ Ⓒ Ⓓ Ⓔ
25. Ⓐ Ⓑ Ⓒ Ⓓ Ⓔ

26. Ⓐ Ⓑ Ⓒ Ⓓ Ⓔ .
27. Ⓐ Ⓑ Ⓒ Ⓓ Ⓔ
28. Ⓐ Ⓑ Ⓒ Ⓓ Ⓔ
29. Ⓐ Ⓑ Ⓒ Ⓓ Ⓔ
30. Ⓐ Ⓑ Ⓒ Ⓓ Ⓔ
31. Ⓐ Ⓑ Ⓒ Ⓓ Ⓔ
32. Ⓐ Ⓑ Ⓒ Ⓓ Ⓔ
33. Ⓐ Ⓑ Ⓒ Ⓓ Ⓔ
34. Ⓐ Ⓑ Ⓒ Ⓓ Ⓔ
35. Ⓐ Ⓑ Ⓒ Ⓓ Ⓔ
36. Ⓐ Ⓑ Ⓒ Ⓓ Ⓔ
37. Ⓐ Ⓑ Ⓒ Ⓓ Ⓔ
38. Ⓐ Ⓑ Ⓒ Ⓓ Ⓔ
39. Ⓐ Ⓑ Ⓒ Ⓓ Ⓔ
40. Ⓐ Ⓑ Ⓒ Ⓓ Ⓔ
41. Ⓐ Ⓑ Ⓒ Ⓓ Ⓔ
42. Ⓐ Ⓑ Ⓒ Ⓓ Ⓔ
43. Ⓐ Ⓑ Ⓒ Ⓓ Ⓔ
44. Ⓐ Ⓑ Ⓒ Ⓓ Ⓔ
45. Ⓐ Ⓑ Ⓒ Ⓓ Ⓔ
46. Ⓐ Ⓑ Ⓒ Ⓓ Ⓔ
47. Ⓐ Ⓑ Ⓒ Ⓓ Ⓔ
48. Ⓐ Ⓑ Ⓒ Ⓓ Ⓔ
49. Ⓐ Ⓑ Ⓒ Ⓓ Ⓔ
50. Ⓐ Ⓑ Ⓒ Ⓓ Ⓔ

SAT Math Level 2

Practice Test 2

1. Ⓐ Ⓑ Ⓒ Ⓓ Ⓔ
2. Ⓐ Ⓑ Ⓒ Ⓓ Ⓔ
3. Ⓐ Ⓑ Ⓒ Ⓓ Ⓔ
4. Ⓐ Ⓑ Ⓒ Ⓓ Ⓔ
5. Ⓐ Ⓑ Ⓒ Ⓓ Ⓔ
6. Ⓐ Ⓑ Ⓒ Ⓓ Ⓔ
7. Ⓐ Ⓑ Ⓒ Ⓓ Ⓔ
8. Ⓐ Ⓑ Ⓒ Ⓓ Ⓔ
9. Ⓐ Ⓑ Ⓒ Ⓓ Ⓔ
10. Ⓐ Ⓑ Ⓒ Ⓓ Ⓔ
11. Ⓐ Ⓑ Ⓒ Ⓓ Ⓔ
12. Ⓐ Ⓑ Ⓒ Ⓓ Ⓔ
13. Ⓐ Ⓑ Ⓒ Ⓓ Ⓔ
14. Ⓐ Ⓑ Ⓒ Ⓓ Ⓔ
15. Ⓐ Ⓑ Ⓒ Ⓓ Ⓔ
16. Ⓐ Ⓑ Ⓒ Ⓓ Ⓔ
17. Ⓐ Ⓑ Ⓒ Ⓓ Ⓔ
18. Ⓐ Ⓑ Ⓒ Ⓓ Ⓔ
19. Ⓐ Ⓑ Ⓒ Ⓓ Ⓔ
20. Ⓐ Ⓑ Ⓒ Ⓓ Ⓔ
21. Ⓐ Ⓑ Ⓒ Ⓓ Ⓔ
22. Ⓐ Ⓑ Ⓒ Ⓓ Ⓔ
23. Ⓐ Ⓑ Ⓒ Ⓓ Ⓔ
24. Ⓐ Ⓑ Ⓒ Ⓓ Ⓔ
25. Ⓐ Ⓑ Ⓒ Ⓓ Ⓔ
26. Ⓐ Ⓑ Ⓒ Ⓓ Ⓔ
27. Ⓐ Ⓑ Ⓒ Ⓓ Ⓔ
28. Ⓐ Ⓑ Ⓒ Ⓓ Ⓔ
29. Ⓐ Ⓑ Ⓒ Ⓓ Ⓔ
30. Ⓐ Ⓑ Ⓒ Ⓓ Ⓔ
31. Ⓐ Ⓑ Ⓒ Ⓓ Ⓔ
32. Ⓐ Ⓑ Ⓒ Ⓓ Ⓔ
33. Ⓐ Ⓑ Ⓒ Ⓓ Ⓔ
34. Ⓐ Ⓑ Ⓒ Ⓓ Ⓔ
35. Ⓐ Ⓑ Ⓒ Ⓓ Ⓔ
36. Ⓐ Ⓑ Ⓒ Ⓓ Ⓔ
37. Ⓐ Ⓑ Ⓒ Ⓓ Ⓔ
38. Ⓐ Ⓑ Ⓒ Ⓓ Ⓔ
39. Ⓐ Ⓑ Ⓒ Ⓓ Ⓔ
40. Ⓐ Ⓑ Ⓒ Ⓓ Ⓔ
41. Ⓐ Ⓑ Ⓒ Ⓓ Ⓔ
42. Ⓐ Ⓑ Ⓒ Ⓓ Ⓔ
43. Ⓐ Ⓑ Ⓒ Ⓓ Ⓔ
44. Ⓐ Ⓑ Ⓒ Ⓓ Ⓔ
45. Ⓐ Ⓑ Ⓒ Ⓓ Ⓔ
46. Ⓐ Ⓑ Ⓒ Ⓓ Ⓔ
47. Ⓐ Ⓑ Ⓒ Ⓓ Ⓔ
48. Ⓐ Ⓑ Ⓒ Ⓓ Ⓔ
49. Ⓐ Ⓑ Ⓒ Ⓓ Ⓔ
50. Ⓐ Ⓑ Ⓒ Ⓓ Ⓔ

SAT Math Level 2

Practice Test 3

1. Ⓐ Ⓑ Ⓒ Ⓓ Ⓔ		26. Ⓐ Ⓑ Ⓒ Ⓓ Ⓔ	
2. Ⓐ Ⓑ Ⓒ Ⓓ Ⓔ		27. Ⓐ Ⓑ Ⓒ Ⓓ Ⓔ	
3. Ⓐ Ⓑ Ⓒ Ⓓ Ⓔ		28. Ⓐ Ⓑ Ⓒ Ⓓ Ⓔ	
4. Ⓐ Ⓑ Ⓒ Ⓓ Ⓔ		29. Ⓐ Ⓑ Ⓒ Ⓓ Ⓔ	
5. Ⓐ Ⓑ Ⓒ Ⓓ Ⓔ		30. Ⓐ Ⓑ Ⓒ Ⓓ Ⓔ	
6. Ⓐ Ⓑ Ⓒ Ⓓ Ⓔ		31. Ⓐ Ⓑ Ⓒ Ⓓ Ⓔ	
7. Ⓐ Ⓑ Ⓒ Ⓓ Ⓔ		32. Ⓐ Ⓑ Ⓒ Ⓓ Ⓔ	
8. Ⓐ Ⓑ Ⓒ Ⓓ Ⓔ		33. Ⓐ Ⓑ Ⓒ Ⓓ Ⓔ	
9. Ⓐ Ⓑ Ⓒ Ⓓ Ⓔ		34. Ⓐ Ⓑ Ⓒ Ⓓ Ⓔ	
10. Ⓐ Ⓑ Ⓒ Ⓓ Ⓔ		35. Ⓐ Ⓑ Ⓒ Ⓓ Ⓔ	
11. Ⓐ Ⓑ Ⓒ Ⓓ Ⓔ		36. Ⓐ Ⓑ Ⓒ Ⓓ Ⓔ	
12. Ⓐ Ⓑ Ⓒ Ⓓ Ⓔ		37. Ⓐ Ⓑ Ⓒ Ⓓ Ⓔ	
13. Ⓐ Ⓑ Ⓒ Ⓓ Ⓔ		38. Ⓐ Ⓑ Ⓒ Ⓓ Ⓔ	
14. Ⓐ Ⓑ Ⓒ Ⓓ Ⓔ		39. Ⓐ Ⓑ Ⓒ Ⓓ Ⓔ	
15. Ⓐ Ⓑ Ⓒ Ⓓ Ⓔ		40. Ⓐ Ⓑ Ⓒ Ⓓ Ⓔ	
16. Ⓐ Ⓑ Ⓒ Ⓓ Ⓔ		41. Ⓐ Ⓑ Ⓒ Ⓓ Ⓔ	
17. Ⓐ Ⓑ Ⓒ Ⓓ Ⓔ		42. Ⓐ Ⓑ Ⓒ Ⓓ Ⓔ	
18. Ⓐ Ⓑ Ⓒ Ⓓ Ⓔ		43. Ⓐ Ⓑ Ⓒ Ⓓ Ⓔ	
19. Ⓐ Ⓑ Ⓒ Ⓓ Ⓔ		44. Ⓐ Ⓑ Ⓒ Ⓓ Ⓔ	
20. Ⓐ Ⓑ Ⓒ Ⓓ Ⓔ		45. Ⓐ Ⓑ Ⓒ Ⓓ Ⓔ	
21. Ⓐ Ⓑ Ⓒ Ⓓ Ⓔ		46. Ⓐ Ⓑ Ⓒ Ⓓ Ⓔ	
22. Ⓐ Ⓑ Ⓒ Ⓓ Ⓔ		47. Ⓐ Ⓑ Ⓒ Ⓓ Ⓔ	
23. Ⓐ Ⓑ Ⓒ Ⓓ Ⓔ		48. Ⓐ Ⓑ Ⓒ Ⓓ Ⓔ	
24. Ⓐ Ⓑ Ⓒ Ⓓ Ⓔ		49. Ⓐ Ⓑ Ⓒ Ⓓ Ⓔ	
25. Ⓐ Ⓑ Ⓒ Ⓓ Ⓔ		50. Ⓐ Ⓑ Ⓒ Ⓓ Ⓔ	

SAT Math Level 2

Practice Test 4

1. Ⓐ Ⓑ Ⓒ Ⓓ Ⓔ
2. Ⓐ Ⓑ Ⓒ Ⓓ Ⓔ
3. Ⓐ Ⓑ Ⓒ Ⓓ Ⓔ
4. Ⓐ Ⓑ Ⓒ Ⓓ Ⓔ
5. Ⓐ Ⓑ Ⓒ Ⓓ Ⓔ
6. Ⓐ Ⓑ Ⓒ Ⓓ Ⓔ
7. Ⓐ Ⓑ Ⓒ Ⓓ Ⓔ
8. Ⓐ Ⓑ Ⓒ Ⓓ Ⓔ
9. Ⓐ Ⓑ Ⓒ Ⓓ Ⓔ
10. Ⓐ Ⓑ Ⓒ Ⓓ Ⓔ
11. Ⓐ Ⓑ Ⓒ Ⓓ Ⓔ
12. Ⓐ Ⓑ Ⓒ Ⓓ Ⓔ
13. Ⓐ Ⓑ Ⓒ Ⓓ Ⓔ
14. Ⓐ Ⓑ Ⓒ Ⓓ Ⓔ
15. Ⓐ Ⓑ Ⓒ Ⓓ Ⓔ
16. Ⓐ Ⓑ Ⓒ Ⓓ Ⓔ
17. Ⓐ Ⓑ Ⓒ Ⓓ Ⓔ
18. Ⓐ Ⓑ Ⓒ Ⓓ Ⓔ
19. Ⓐ Ⓑ Ⓒ Ⓓ Ⓔ
20. Ⓐ Ⓑ Ⓒ Ⓓ Ⓔ
21. Ⓐ Ⓑ Ⓒ Ⓓ Ⓔ
22. Ⓐ Ⓑ Ⓒ Ⓓ Ⓔ
23. Ⓐ Ⓑ Ⓒ Ⓓ Ⓔ
24. Ⓐ Ⓑ Ⓒ Ⓓ Ⓔ
25. Ⓐ Ⓑ Ⓒ Ⓓ Ⓔ
26. Ⓐ Ⓑ Ⓒ Ⓓ Ⓔ
27. Ⓐ Ⓑ Ⓒ Ⓓ Ⓔ
28. Ⓐ Ⓑ Ⓒ Ⓓ Ⓔ
29. Ⓐ Ⓑ Ⓒ Ⓓ Ⓔ
30. Ⓐ Ⓑ Ⓒ Ⓓ Ⓔ
31. Ⓐ Ⓑ Ⓒ Ⓓ Ⓔ
32. Ⓐ Ⓑ Ⓒ Ⓓ Ⓔ
33. Ⓐ Ⓑ Ⓒ Ⓓ Ⓔ
34. Ⓐ Ⓑ Ⓒ Ⓓ Ⓔ
35. Ⓐ Ⓑ Ⓒ Ⓓ Ⓔ
36. Ⓐ Ⓑ Ⓒ Ⓓ Ⓔ
37. Ⓐ Ⓑ Ⓒ Ⓓ Ⓔ
38. Ⓐ Ⓑ Ⓒ Ⓓ Ⓔ
39. Ⓐ Ⓑ Ⓒ Ⓓ Ⓔ
40. Ⓐ Ⓑ Ⓒ Ⓓ Ⓔ
41. Ⓐ Ⓑ Ⓒ Ⓓ Ⓔ
42. Ⓐ Ⓑ Ⓒ Ⓓ Ⓔ
43. Ⓐ Ⓑ Ⓒ Ⓓ Ⓔ
44. Ⓐ Ⓑ Ⓒ Ⓓ Ⓔ
45. Ⓐ Ⓑ Ⓒ Ⓓ Ⓔ
46. Ⓐ Ⓑ Ⓒ Ⓓ Ⓔ
47. Ⓐ Ⓑ Ⓒ Ⓓ Ⓔ
48. Ⓐ Ⓑ Ⓒ Ⓓ Ⓔ
49. Ⓐ Ⓑ Ⓒ Ⓓ Ⓔ
50. Ⓐ Ⓑ Ⓒ Ⓓ Ⓔ